The Handbook
of Environmental Chemistry

Volume 3   Part B

Edited by O. Hutzinger

Anthropogenic Compounds

With Contributions by
K. J. Bock, K. A. Daum, E. Merian, L. W. Newland,
C. R. Pearson, H. Stache, M. Zander

With 38 Figures

Springer-Verlag Berlin Heidelberg GmbH 1982

Professor Dr. Otto Hutzinger
Laboratory of Environmental and Toxicological Chemistry
University of Amsterdam, Nieuwe Achtergracht 166
Amsterdam, The Netherlands

ISBN 978-3-662-15334-5

Library of Congress Cataloging in Publication Data
Main entry under title: Anthropogenic compounds. (The Handbook of environmental chemistry; v. 3, pt. A.–B.)
Includes bibliographies and index.
1. Pollution – Environmental aspects. 2. Pollution – Toxicology. 3. Environmental chemistry.
I. Butler, Gordon Cecil, 1913–. II. Bock, K. J. III. Series: Handbook of environmental chemistry; v. 3, pt. A.–B.
QD31.H335 vol. 3, pt. A, etc. 574.5′222s   80-16609
[QH545.A1] [574.5′222]   AACR1
ISBN 978-3-662-15334-5   ISBN 978-3-540-47028-1 (eBook)
DOI 10.1007/978-3-540-47028-1
Originally published by Springer-Verlag Berlin Heidelberg New York in 1982
Softcover reprint of the hardcover 1st edition 1982

2152/3140-543210

Preface

Environmental Chemistry is a relatively young science. Interest in this subject, however, is growing very rapidly and, although no agreement has been reached as yet about the exact content and limits of this interdisciplinary discipline, there appears to be increasing interest in seeing environmental topics which are based on chemistry embodied in this subject. One of the first objectives of Environmental Chemistry must be the study of the environment and of natural chemical processes which occur in the environment. A major purpose of this series on Environmental Chemistry, therefore, is to present a reasonably uniform view of various aspects of the chemistry of the environment and chemical reactions occurring in the environment.

The industrial activities of man have given a new dimension to Environmental Chemistry. We have now synthesized and described over five million chemical compounds and chemical industry produces about hundred and fifty million tons of synthetic chemicals annually. We ship billions of tons of oil per year and through mining operations and other geophysical modifications, large quantities of inorganic and organic materials are released from their natural deposits. Cities and metropolitan areas of up to 15 million inhabitants produce large quantities of waste in relatively small and confined areas. Much of the chemical products and waste products of modern society are released into the environment either during production, storage, transport, use or ultimate disposal. These released materials participate in natural cycles and reactions and frequently lead to interference and disturbance of natural systems.

Environmental Chemistry is concerned with *reactions in the environment*. It is about distribution and equilibria between environmental compartments. It is about reactions, pathways, thermodynamics and kinetics. An important purpose of this Handbook is to aid understanding of the basic distribution and chemical reaction processes which occur in the environment.

Laws regulating toxic substances in various countries are designed to assess and control risk of chemicals to man and his environment. Science can contribute in two areas to this assessment; firstly in the area of toxicology and secondly in the area of chemical exposure. The available concentration ("environmental exposure concentration") depends on the fate of chemical compounds in the environment and thus their distribution and reaction behaviour in the environment. One very important contribution of Environmental Chemistry to the above mentioned toxic substances laws is to develop laboratory test methods, or mathematical correlations and models that predict the environmental fate of new chemical compounds. The third purpose of this Handbook is to help in the basic understanding and development of such test methods and models.

The last explicit purpose of the Handbook is to present, in concise form, the most important properties relating to environmental chemistry and hazard assessment for the most important series of chemical compounds.

At the moment three volumes of the Handbook are planned. Volume 1 deals with the natural environment and the biogeochemical cycles therein, including some background information such as energetics and ecology. Volume 2 is concerned with reactions and processes in the environment and deals with physical factors such as transport and adsorption, and chemical, photochemical and biochemical reactions in the environment, as well as some aspects of pharmacokinetics and metabolism within organisms. Volume 3 deals with anthropogenic compounds, their chemical backgrounds, production methods and information about their use, their environmental behaviour, analytical methodology and some important aspects of their toxic effects. The material for volume 1, 2 and 3 was each more than could easily be fitted into a single volume, and for this reason, as well as for the purpose of rapid publication of available manuscripts, all three volumes were divided in the parts A and B. Publisher and editor hope to keep materials of the volumes one to three up to date and to extend coverage in the subject areas by publishing further parts in the future. Readers are encouraged to offer suggestions and advice as to future editions of "The Handbook of Environmental Chemistry".

Most chapters in the Handbook are written to a fairly advanced level and should be of interest to the graduate student and practising scientist. I also hope that the subject matter treated will be of interest to people outside chemistry and to scientists in industry as well as government and regulatory bodies. It would be very satisfying for me to see the books used as a basis for developing graduate courses on Environmental Chemistry.

Due to the breadth of the subject matter, it was not easy to edit this Handbook. Specialists had to be found in quite different areas of science who were willing to contribute a chapter within the prescribed schedule. It is with great satisfaction that I thank all 52 authors from 8 countries for their understanding and for devoting their time to this effort. Special thanks are due to Dr. F. Boschke of Springer for his advice and discussions throughout all stages of preparation of the Handbook. Mrs. A. Heinrich of Springer has significantly contributed to the technical development of the book through her conscientious and efficient work. Finally I like to thank my family, students and colleagues for being so patient with me during several critical phases of preparation for the Handbook, and to some colleagues and the secretaries for technical help.

I consider it a privilege to see my chosen subject grow. My interest in Environmental Chemistry dates back to my early college days in Vienna. I received significant impulses during my postdoctoral period at the University of California and my interest slowly developed during my time with the

National Research Council of Canada, before I could devote my full time to Environmental Chemistry, here in Amsterdam. I hope this Handbook may help deepen the interest of other scientists in this subject.

O. Hutzinger

Contents

Lead

L. W. Newland and K. A. Daum

Arsenic, Beryllium, Selenium and Vanadium

L. W. Newland

C$_1$ and C$_2$ Halocarbons

C. R. Pearson

Halogenated Aromatics

C. R. Pearson

Volatile Aromatics

E. Merian and M. Zander

Surfactants

K. J. Bock and H. Stache

Volume 1, Part B: **The Natural Environment and the Biogeochemical Cycles**

Volume 2, Part B: **Reactions and Processes**

List of Contributors

Dr. K. J. Bock
Dr. H. Stache
Chemische Werke Hüls AG
Postfach 1320
D-4370 Marl
Federal Republic of Germany

Dr. K. A. Daum
Operations Analysis Division
Research Triangle Institute
P.O.B. 12194
Research Triangle Park
NC 27709, USA

Dr. E. Merian
International Association
of Environmental Analytical Chemistry
and Swiss Association
for Environmental Research
Im Kirsgarten 22
CH-4106 Therwil, Switzerland

Dr. L. W. Newland
Environmental Sciences Program
Texas Christian University
Fort Worth, TX 76129, USA

Dr. C. R. Pearson
Imperial Chemical Industries Ltd.
Brixham Laboratory
Freshwater Quarry
Overgang Brixham,
Devon TQ5 8BA, U.K.

Prof. Dr. M. Zander
Rütgerswerke AG
Postfach 504
Kekuléstraße 30
D-4620 Castrop-Rauxel
Federal Republic of Germany

Lead

L. W. Newland

Texas Christian University
Fort Worth, Texas, USA

K. A. Daum

Research Triangle Institute
Research Triangle Park, North Carolina, USA

Introduction

The quantity of published data available on the pollution chemistry of lead is staggering. Books and journal articles on lead are probably as numerous as those for any other pollutant. Because of the ease with which lead is mined and smelted, it was one of the first metals used by early man. Lead was used for glazing pottery, and making ornaments by the Egyptians in 7,000–5,000 BC. The Romans used lead pots for cooking, lead salts for sweeteners, and lead plumbing in their cities. There are even references to the discovery of the toxicological properties of lead by the Romans and the Greeks. High lead concentrations in the bones of Roman aristocrats have led some to believe that lead poisoning advanced the fall of the Roman Empire [50]. Since the technological revolution of the 18th and 19th centuries a more enlightened approach to public health problems has brought about extensive study of lead. It is hoped that this chapter will present some introduction to the many aspects of lead pollution.

Production, Use and Natural Occurrence

Production

Lead Mining

Lead occurs as a variety of ores, the most important of which is galena, with cerussite and anglestie being of secondary importance. Sphalerite (zinc), chalcopyrite (copper), and silver are common coproducts commercially mined with lead. Other minor constituents of lead ores are gold, bismuth, antimony, arsenic, cadmium, tin, gallium, thallium, indium, germanium, and tellurium [77].

In 1965, the world mine production of lead was 2.6 million Mg, with production increasing to 3.6 million Mg in 1975. The most important lead mining countries in 1975 were; the United States (16.0% total world output), the Soviet Union (14.5%), Australia (10.0%), Canada (9.6%), Peru (5.5%), Mexico (4.5%), China (3.8%), Yugoslavia (3.5%), and Bulgaria (3.0%). In addition, Ireland, Japan, the Democratic People's Republic of Korea, Morocco, Poland, Spain, and Sweden each had over 2% of the total world production of lead. The estimated proven lead reserves of the world are 93 million Mg by metal content.

Smelting and Refining

In the primary smelting and refining of lead, concentrated metallic minerals from the mine are formed into pellets, which are roasted. In the case of the primary lead ore, galena (PbS), the roasting process removes the S, creating a sinter that is fired along with coke to chemically reduce the Pb to its metallic form. The lead concentrate then undergoes a high-pressure aqueous oxidation to remove other metals that hinder the smelting process. The secondary smelting and refining of lead uses new process scrap that comes from manufacturing processes and old recycled scrap that comes from discarded lead-containing manufactured goods [45, 77].

Environmental Pollution from Production

Anthropogenic sources of lead in the environment from the production of lead can be associated with (a) the mining-milling operations, which include grinding, concentrating, and transportation of the ore, as well as disposal of the tailings and mine and mill wastewater and (b) the smelter-refininery process and problems associated with concentrate hauling, storage, sintering, refining, atmospheric discharges, and blowing dust [76].

In 1975, there was an estimated total of 19,225 Mg of lead emitted to the atmosphere from stationary sources. Of this total, 400 Mg were from primary lead smelting, and 755 Mg were from secondary lead smelting, representing 2.1 and 3.9%, respectively, of the total lead emissions from stationary sources. In 1975, 142,000 Mg of lead were emitted from mobil sources through the combustion of gasoline [73].

Use

Industry

Lead pigments are commonly used in paints, although less toxic pigments are presently used preferentially. Red lead (minimum) is used extensively in the painting of structural steel, and lead chromate is used as a yellow pigment (see Table 1). At

Table 1. Percentage of total lead consumption in 1969 and 1974 for selected industries in major industrial countries

Industry	1969	1974
Chemical	10.9	12.0
Cable and sheathing	10.9	9.2
Storage batteries	35.9	44.0
Lead in fuels	12.0	12.0
Alloys	8.1	10.8
Semi-Manufacturers	16.5	12.0

Source: WHO, 1977 [77]

one time, lead arsenate was extensively used as an insecticide, although present consumption is considerably reduced. Litharge (PbO) dissolved in a sodium hydroxide solution is used to remove sulfur compounds from petroleum during the refining process [77].

Lead consumption in the cable industry has declined because of the introduction of plastic sheathing and insulation (see Table 1). However, the total amount of lead used in the industry is significant. Cadmium, tellurium, copper, antimony, and arsenic are trace contaminants in alloys used for cable sheathing [77].

The largest consumer of lead is the electric manufacturing industry storage battery (see Table 1). A lead-antimony alloy is used in the preparation of grids and lugs. Litharge, red lead, and grey oxide (PbO_2) are used in the preparation of pasted plates [45]. The percentage of total lead consumption accounted for by the battery industry is increasing (see Table 1), although the demand for lead batteries has decreased. This may be partially attributed to the increased use of long-life batteries. About 80% of the lead in storage batteries is recovered at secondary smelters, making the battery industry the source for secondary lead production [77].

Lead additives are used in gasoline to increase the octane rating. The emission of these additives through the exhaust of internal combustion engines is the largest source of lead in the atmosphere [73]. The additives are almost exclusively tetramethyllead and tetraethyllead. World consumption of refined lead for the manufacture of lead additives reached a maximum of 380,000 Mg in 1973, with consumption declining 30% by 1975 [33, 77]. This decrease in consumption can be attributed to increased use of catalytic converters on motor vehicles, requiring the use of lead-free gasoline.

The production of manufactured lead components is responsible for an important part of total lead consumption. Lead surfaces oxidize readily, leaving the surface resistant to corrosion. Lead is used in construction when corrosion resistance is desired. Roofing, flashing, wall cladding, and sound insulation are all instances where lead is used. Alloys of lead are used in solder, ball bearings, brasses, typesetting metal, collapsible tubes, and radiation shielding. The manufacture of ammunition is also a major consumer of lead. There are many additional uses of lead components, but these account for a small portion of total lead consumption [77].

Environmental Pollution from Consumption and Uses of Lead

In the early part of this century, smelting and refining of lead were the main sources of lead emissions. As discussed, the combustion of lead in fuel is presently the greatest contributer to lead emissions. Of the emissions, over 70% will enter the environment upon combustion. The remainder is trapped in the crank case oil and in the exhaust system of the vehicles [77]. Pathways and cycling of these emissions are discussed later.

In the United States, 1,900 Mg (11%) of atmospheric lead emissions from industrial sources are attributable to the production of fuel additives, 480 Mg to the manufacture of batteries, and smaller amounts to the manufacture of lead oxide, lead pigments, metal components, and solder [15].

Natural Occurrence

Table 2 presents data on the natural occurrence and concentration of lead in the environment with some references being provided for each major occurrence category.

Chemistry

Elemental Lead, General Inorganic Chemistry

Lead is the most massive of the Group IV B elements, which include carbon, silican, germanium, and tin. As with other elements in its row, the electronic structure of lead is expressed in a filled configuration of xenon, plus the additional partially filled subshells, i.e., [Xe] $4f^{14}$ $5d^{10}$ $6S^2$ $6p^2$. The physical properties of elemental lead are listed in Table 3.

The inorganic chemistry of lead is dominated by the divalent (2^+) oxidation state rather than the tetravalent (4^+) oxidation state. The divalent state is more dominant in Group IV B elements as the atomic number increases. Dominance of the divalent state occurs because, within Group IV B, there is a decrease in single bond strength with increasing atomic number [13].

The average energy of a C-H bond is 100 kcal/mole; it is this factor that allows CH_4 to be more stable than CH_2. For lead, the Pb-H bond energy is only about 65 kcal/mole. This energy is too small to compensate for the Pb(II)→Pb(IV) promotional energy; therefore, PbX_2 compounds are more stable.

As with Si, Ge, and Sn, lead does not form $\varrho\pi$ multiple bonds either with itself or with other elements [13]. This accounts for the lack of lead analogs for the numerous classes of organic compounds.

Table 4 presents solubilities and other physical properties of inorganic lead compounds. Inorganic lead (II) salts of lead sulfide, and lead oxides are poorly soluble in water. Notable exceptions are lead nitrate, lead acetate, lead chlorate, and, to a lesser degree, lead chloride. Inorganic lead (II) salts have relatively high melting points, with corresponding low vapor pressures at room temperature.

Table 2. Environmental inventory of lead

Occurrence	Concentration (ppm)	Reference
Meteorites		Nriagu [50]
Ordinary chondrites	0.45	
Carbonaceous chondrites	1.95	
Enstate chondrites	2.82	
Achondrites	0.44	
Troilite	5.9	
Octahedrites	0.17	
Hexahedrites	0.93	
Igneous rocks		Nriagu [50]
Peridotites	0.2	
Gabbro	1.9	
Tholeitic basalts	3.0	
Olivine basalts	3.8	
Spilite	3.6	
Diorites	9.3	
Andesites	8.3	
Dacite	10.7	
Trachyte	11.9	
Syemite	13.9	
Phonolite	14.6	
Granodiorites	18.9	
Ryolites	21.5	
Granites	22.7	
Metamorphic rocks		Nriagu [50]
Eclogites	1.5	
Gneisses and schists	17.6	
Granulites	18.5	
Marble	3.9	
Hornfels	17.9	
Slates	21.0	
Sedimentary rocks		Nriagu [50]
Sandstones	9.8	
Greywackes	18.2	
Shales, non-bituminous	23.1	
Shales, bituminous	27.4	
Shales, kaolinitic	49.0	
Limestones and dolomites	6.7	
Evaporites	~0.2	
Phosphorites	~10	
Bauxites	>100	
Soils	2–300	Bowen [8]
Dry plant tissues		Bowen [8]
Plankton	4,000–8,000	
Brown algae	2.0–38.0	
Bryophytes	8.0–25.0	
Ferns	2.3	
Gymnosperms	0.9–13.0	
Angiosperms	1.0–8.0	
Fungi	0.2–40.0	

Table 2 (continued)

Occurrence	Concentration (ppm)	Reference
Dry animal tissue		Bowen [8]
Coelenterata	3.0–24	
Mollusca	0.5–42	
Echinodermata	0.5– 7.0	
Crustacea	0.3–49.0	
Pisces	0.001–15.0	
Zooplankton	2.0–130.0	
Annelida	5.0–45.0	
Hard tissues of marine biota		Bowen [8]
Foraminifera ($CaCO_3$)	20.0–140.0	
Foraminifera (SiO_2)	10.0	
Porifera (SiO_2)	5.5	
Echinoderms ($CaCO_3$)	5.0	
Corrals ($CaCO_3$)	2.0	
Dry human tissues		Bowen [7, 8]
Bone	3.6–30.0	
Brain	0.24	
Heart	0.2	
Kidney	1.2– 6.8	
Liver	3.0–12.0	
Lung	2.3	
Muscle	0.2– 3.3	
Skin	0.78	
Hair	3.0–70.0	
Nail	14.0–170.0	
Human blood components		Bowen [8]
Whole blood	0.21	
Plasma	0.13	
Red cells	0.46	
Serum	0.16–0.31	
Atmospheric particulates		EPA [73]
(≤ 0.5 μm diameter)		
Urban (USA)	0.1 –5.0 ng/m^3	
Rural (USA)	0.01–1.4 ng/m^3	
Natural waters		Chow [11]
Rain water	6.2 to over 300	
Snow and ice (modern)	Undetectable–1.090	
Surface waters	0.06–120.0	

Table 3. Physical properties of elemental lead

Property	Value
Atomic weight	207.19
Atomic number	82
Oxidation states	$^+2, ^+4$
Density	11.35 g/cm^3 at 20 °C
Melting point	327.4 °C
Boiling point	1744 °C
Covalent radius (tetrahedral)	1.44 Å
Ionic radii	1.21 Å ($^+2$), 0.84 Å ($^+4$)
Resistivity	21.9×10^{-6} ohm/cm
Electronegativity	1.55
Ionization potentials	
First	7.4 eV
Second	15.0 eV
Third	32.0 eV
Fourth	42.3 eV
Naturally occurring isotopes	
(in order of abundance)	208, 206, 207, and 204

Source: Handbook of Chemistry and Physics, 55th Ed. [29]

Organometallic Chemistry

An organolead compound contains at least one covalent bord between lead and carbon. The divalent lead salts of organic acids, e.g., lead acetate $(CH_3COO)_2Pb$, do not contain a covalent bord between lead and carbon but have an ionic bond between the negatively charged acetate ion, CH_3COO^-, and the positively charged metallic ion, Pb^{2+} [26]. The Pb-C bond energy is about 130 kcal/mole, or about twice the Pb-H value cited above in the previous section. It is for this reason that the organometallic chemistry of lead is dominated by the tetravalent (4^+) oxidation state. Tetravalent organolead compounds, although more stable than most divalent organolead compounds, are not as stable as the inorganic lead (II) salts to which they may be degraded. Also, most of organolead compounds undergo photolysis when exposed to sunlight [73].

The tetraalkyl compounds tetraethyl- and tetramethyl-lead are the most important organolead compounds. This is because of their use as antiknock compounds in gasoline, leading to their abundant and wide distribution in the environment [73]. Tetraalkyllead compounds, R_4Pb, are almost insoluble in water, are stable in air, and are very soluble in nonpolar organic solvents. Tetraethyl- and tetramethyl-lead are clear, colorless liquids having a relatively high vapor pressure at room temperature and a fruity smell. For comparison, most trialkyllead compounds, R_3PbX, are white solids that are stable in air and soluble in water, alcohol, and most organic solvents. Dialkyllead compounds, R_2PbX_2, are also white solids and, because of their greater ionic properties, are more soluble in polar solvents than nonpolar solvents [26].

Tetraethyl- and tetramethyl-lead are removed from internal combustion engines by scavenging, i.e., a reaction with halogenated hydrocarbon gasoline ad-

Table 4. Solubility of some inorganic lead compounds

Compound	Formula	Molecular weight	Solubility, g/100 ml		Other solvents
			Cold water	Hot water	
Lead	Pb	207.19	i	i	sa
Acetate	$Pb(C_2H_3O_2)_2$	325.28	44.3	221	s glyc
Azide	$Pb(N_3)_2$	291.23	0.023	0.09	—
Bromate	$Pb(BrO_3 \cdot H_2O$	481.02	1.38	sls	—
Bromide	$PbBr_2$	367.01	0.8441	4.71	sa
Carbonate	$PbCO_3$	267.20	0.00011	d	sa, alk
Chloride	$PbCl_2$	278.10	0.99	3.34	i al
Chlorobromide	PbClBr	322.56	nd	nd	—
Chromate	$PbCrO_4$	323.18	6×10^{-6}	i	sa, alk
Fluoride	PbF_2	245.19	0.064	nd	s HNO_3
Fluorochloride	PbFCl	261.64	0.037	0.1081	—
Hydride	PbH_2	209.21	nd	nd	—
Hydroxide	$Pb(OH)_2$	241.20	0.0155	sls	sa, alk
Iodate	$Pb(IO_3)_2$	557.00	0.0012	0.003	s HNO_3
Iodide	PbI_2	461.00	0.063	0.41	s, alk
Nitrate	$Pb(NO_3)_2$	331.20	37.65	127	s, alk
Oxalate	PbC_2O_4	295.21	0.00016	nd	sa
Oxide	PbO	223.19	0.0017	nd	s, alk
di Oxide	PbO_2	239.19	i	i	sa
Oxide (red)	Pb_3O_4	685.57	i	i	sa
Phosphate	$Pb_3(PO_4)_2$	811.51	1.4×10^{-5}	i	s, alk
Sulfate	$PbSO_4$	303.25	0.00425	0.0056	—
Sulfide	PbS	239.25	8.6×10^{-5}	nd	sa
Sulfite	$PbSO_3$	287.25	i	i	sa
Thiocyanate	$Pb(SCN)_2$	323.35	0.05	0.2	s, alk

Abbreviations:
a   = acid
al   = alcohol
alk = alkali
d   = decomposes
expl = explodes
glyc = glycol
i   = insoluble
s   = soluble
nd   = no data
Source: Handbook of Chemistry and Physics, 55 th Edition [29].

ditives (e.g., ethylene dibromide and ethylene dichloride) to form lead halides. These lead halides originate as vapors in the combustion chamber but condense to form microscopic spherical particles during exhaust. The scavenged lead can also condense or adsorb onto the surfaces of co-entrained particles. Consequently, lead halides emitted from automobile exhausts are present as vapors, as pure solid particles, and as a coating on the surface of particulate substrates [73].

An excellent survey of the chemical and physical properties and the preparation of organolead compounds has been prepared by Shapiro and Frey [62].

Complex Formation and Chelation

A metal ion may combine with a negative ion, or a neutral compound, to form a complex or coordination compound. A chelating agent has more than one atom that may be bonded to a single central metal ion to form a ring structure [46]. The bonding in organometallic derivatives of lead is principally covalent rather than ionic because of the similar electronegativities of lead and carbon. The bonding in organolead compounds is the donar-acceptor type, in which both electrons in the bonding orbital are donated by the carbon atom. The doner atoms in a metal complex are usually called ligands and must have a pair of electrons available for bond formation. In general, since a chelating agent may bond to a metal ion in more than one place simultaneously, chelates are more stable than complexes of unidentate ligands.

In simple complexes of lead with unidentate ligands, lead is usually in the Pb(IV) form. In chelated lead compounds with polydentate ligands, lead is usually in the Pb(II) form. These chelates, usually six-coordinate, are kinetically quite labile, although highly thermodynamically stable [73].

Geochemistry

The composition of the soil and the availability of elements from it to plants are determined by the parent rock type, climatic actions, living organisms in the soil, soil management practices, and environmental pollution. The specific chemistry of lead in the soil is affected by
(a) adsorption on mineral surfaces,
(b) the formation of stable chelates with soil organic matter, and
(c) the precipitation of sparingly soluble lead compounds.

Analytical Methods

Sampling Methods

The purpose of sampling is to obtain lead-containing particles, adsorbed gases, liquids, and solid samples that will indicate the spacial, temporal, and chemical nature and the concentration of lead in the environment. Method of sample collection, sampling site selection, and sample processing procedures are all of major importance in sampling methods for lead.

In air sampling, high-volume samplers are preferable for accuracy, but low-volume techniques are useful for obtaining extensive data [77]. More informative data can be obtained if
(1) the required amount of particulates is estimated before the sample volume and procedure are selected,
(2) the sampling devices are positioned in appropriate places to capture representative samples, and
(3) the sample is taken of an appropriate rate and volume and for a sufficient time to obtain an accurate estimate of average concentration.

Techniques for sampling water are less complex than those for sampling air [77]. Filtering is an important consideration in water analysis, since lead can occur in the particulate fraction or in solution in aqueous samples. The preparation of soil and soil dust samples for lead analysis usually involves drying, homogenation by grinding, and sieving to obtain a particle size distribution.

Analysis Methods

Colorimetric Analysis

The reference method most commonly used to test other analysis methods and historically used to analyze lead in samples is the dithizone complexiometric method. In this method, a dithizone-lead complex is formed, which has a broad and intense absorption peak at 510 nm [53]. Quantitative, precise results can be obtained when care is taken in sample and reference preparation. Advantages of the dithizone complexiometric method include simple and relatively inexpensive equipment, linear absorption to lead concentration, need for only a small sample, adaptability to large samples, and ready removal of interferances [73].

Atomic Absorption Analysis

Atomic absorption (AA) spectrometry is a generally accepted method for the analysis of many metals [74]. In a typical AA method, a liquid sample is aspirated into a flame, where ions within the liquid are reduced to the atomic state. The metals in the atomic state can then quantitatively absorb light at the wavelengths characteristic of their resonance frequencies, 217.0 and 283.3 nm for lead. Alternately, the ions may either be chemically reduced by a cold vapor technique [32] or be thermally reduced in a graphite furnace before analysis, which usually gives better sensitivities than flame AA (as low as 50 µg/kg).

Several hundred samples can be analyzed in a work day if the samples are already prepared. In flame techniques and cold vapor techniques, the sample must be in solution. Procedures for solution of some samples may be time consuming. Use of a graphite furnace for sample reduction eliminates many problems in sample preparation.

Anodic Stripping Voltammetry

Anodic stripping voltammetry is an electroanalytical analysis method, that can be employed in the study of many metals [74]. Lead is analyzed by selective deposition on an electrode (reduction) to facilitate concentration. The lead is then stripped (oxidized) by a linearly variable applied voltage, with the output being a plot of current and voltage in which peak area corresponds to the oxidation of lead. The method is accurate and reliable to the 1 µg/L level [73].

Emission Spectroscopy

Optical emission spectroscopy includes the observation of flame-, arc-, and spark-induced emission phenomena in the ultraviolet, visible, and near infrared regions of the electromagnetic spectrum [38]. Qualitative and quantitative information can be gained from the intensity of the characteristic emission wavelengths. Analysis of lead in environmental samples (e.g., soils, rocks, and minerals) may be performed reproducibly down to the 5 ppm level. Emission spectroscopy is best used for the multi-elemental analysis of samples, because of the high cost of equipment. Usually, single element analyses are not performed on a emission spectrograph.

Electron Microprobe

An electron microprobe measures the X-rays produced from the incidence of a beam of electrons on a material. The wavelengths of the X-rays are characteristic of each element, and the intensities of the X-rays depend on the quantity of the element. The technique is accurate to 1%–3%, with the allowable mass of the element analyzed in the sample ranging from 10^{-14} to 10^{-16} g [73].

From an environmental monitoring viewpoint, this method has been used to determine the elemental composition of complex lead particulates [70]. However, the equipment for electron microprobe analysis is expensive, and sample preparation is complex; hence, the method is not extensively used.

X-Ray Fluorescence

Characteristic X-ray spectra are emitted when an element is irradiated with a beam of X-radiation, of a sufficiently short wavelength [74]. This fluorescence allows simultaneous identification of a range of elements, including lead. The technique is identical to electron microprobe analysis, except that the excitation source is different. A high energy source is needed, with X-ray tubes, electron beams, and radioactive sources being commonly used. More recently, charged-particle excitation (proton-induced X-ray emission, or PIXE) has been used [35]. X-ray fluorescence requires minimal sample preparation, can detect a variety of elements, and is available in fully automated formats [73]. There are some matrix effects for the analysis of lead in some complex samples, and the equipment is expensive. However, the method is used extensively to survey dwellings for hazardous concentrations of lead [77].

Compound Analysis

The previously discussed analysis methods give information only on the total lead content of the analyte. Electron microprobe and X-ray fluorescene analysis can provide some information on the molecular nature of the lead compound, by comparing the ratios of the elements present [70]. There are, however, several tools that the analyst may use to identify the molecular nature of lead compounds more com-pletely. These include infrared spectrophotometry, X-ray diffraction, and chroma-tography.

Infrared (IR) spectrophotometric analysis is based on the fact that the charac-teristic wavelengths at which radiation is absorbed or emitted by minerals in the IR region can be related to the inter-atomic vibrations in the molecules or crysals [72]. No two minerals give exactly the same pattern when transmission of radiation is plotted against wavelength. By combining quantitative and qualitative capa-bilities, IR spectrophotometry provides a powerful tool for crystal chemical studies. Cerussite ($PbCO_3$) was one of several carbonates studied by Alder and Kerr [1].

X-ray diffraction is a qualitative and quantitative method for the analysis of minerals [43]. In this method, the repeating interplanar spacings in a crystal are measured by determining the angles of X-rays reflected from the crystal planes. Ol-son and Skogerboe [51] have used X-ray diffraction in the identification of soil lead compounds from automotive sources.

Chromatography is a compound separation method based on the differences in rates at which the individual components of a mixture migrate through a station-ary medium under the influence of a moving phase. The use of gas-liquid chroma-tography with an electron capture detector has been demonstrated in the analysis of organolead compounds [62].

Transport Behavior in the Environment

Reservoirs, Pathways, and Cycling

Figure 1 is a simplified diagram of the ecological cycle of lead. The ecological lead cycle contains reservoirs, between which pathways exist for the transfer of lead. The capacity of a reservoir is the product of the average lead concentration and the total mass. The majority of the lead in the earth's crust resides in the rocks and sediments. Although the atmospheric lead pool is small, it is the most important reservoir, as it accounts for the vast majority of the transfer of lead between other reservoirs. Data on movement of lead within and among the various environmen-tal media are usually only semiquantitative because physical and chemical trans-formation occurring within the cycle are not completely understood [73].

Transport in Air

From a mass-balance point of view, the transport and distribution of lead from sta-tionary and mobile sources into other environmental media are mainly through the atmosphere. The mechanisms of atmospheric transport of gases and particulates are complex. Atmospheric lead emissions are primarily inorganic particulates whose cycling depends on particle size, chemical stability, height of injection, and local atmospheric motions. Therefore, large particles injected at low elevations will settle to the surface in the immediate vicinity of the source, while smaller particles and those injected at higher elevations will be transported over greater distances [73].

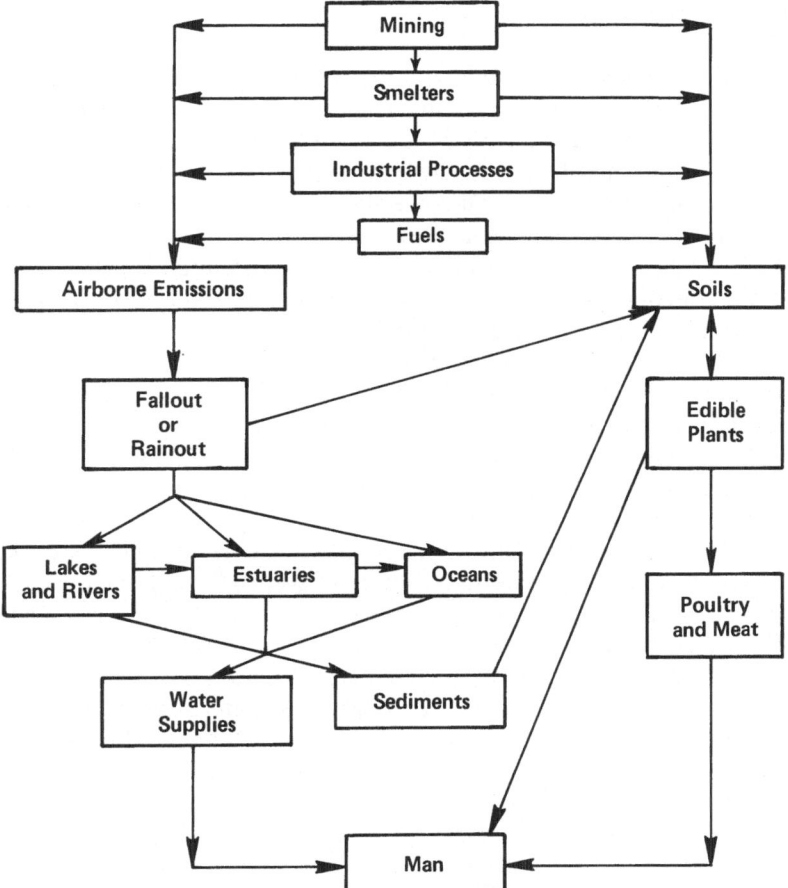

Fig. 1. Ecological cycle of lead

There is a direct relationship between volume of traffic and the lead content of the roadside air [14]. Cholak et al. [10] and Schuck and Locke [61] have found a direct relationship between particle size and distance of deposition from the road-side. About 65 percent of the lead in the air between 9 m and 533 m from the high-way consist of particles with a diameter less than 2 µm, and 85% of the particles have diameters less than 4 µm. Particles greater than 4 µm are severely attenuated after emission from a vehicle.

The principal mechanisms for the removal of inorganic lead particulates from the atmosphere are dry and wet deposition [54]. The removal efficiency depends on the physical characteristics of the suspended material, atmospheric conditions, and the nature of the receiving surface. Dry deposition occurs by sedimentation, diffusion, and inertial mechanisms such as impaction [73]. Wet deposition removal processes include rainout and washout [71].

Transport in Water

In aqueous systems, the inorganic and organic compounds of lead have unique chemistries and transport mechanisms. Normally, naturally occuring lead in ore deposits do not move appreciably in the environment.

The compounds usually formed by lead and the major anions have low solubilities in ground or surface water. Dissolved lead from lead sulfide ore (galena) tends to from insoluble lead carbonate and lead sulfate, or be adsorbed by ferric hydroxide [44]. The quantity of lead available to remain in solution is dependent on the pH, Eh, and dissolved salt content of the water. Stumm and Morgan [67] and Garrels and Christ [23] present detailed descriptions and diagrams for lead in various aqueous systems. Under anaerobic reducing conditions insoluble solid PbS predominates, while in aerobic conditions S^{2-} is oxidized to SO_4^{2-}, limiting the lead concentration in solution. Above pH 5.4, $PbCO_3$ and $Pb_2(OH)_2CO_3$ limits the concentration. The fraction of unbound lead in solution decreases as pH increases, in oxidizing conditions. In general, the unbound fractions are larger oxidizing than for reducing conditions.

In lotic (running water) conditions, lead may be carried in an undissolved state. This undissolved inorganic lead may consist of suspended colloidal particles and larger undissolved particles of lead carbonate, oxide, and hydroxide. The undissolved lead may also be transported by sorption on mineral particles or carried as part of the suspended living or nonliving organic matter [44].

The naturally occurring organic components of water systems include; humic materials, amino acids, carbohydrates, phenolic and quinonoid compounds, organic acids, nucleic acids, enzymes, and porphyrins [44]. In addition, natural waters contain organics and suspended solids from municipal, agricultural, and industrial wastes. There is some question as to the interaction of lead with these compounds to form organometallic complexes, although it is assumed that complexes will be formed because of the availability of donar sites. Stumm and Bilinski [66] considered that trace metal chelates are found only in those natural waters where organic substances become enriched, although hydroxide and carbonate complexes of lead may still be the predominant species even if organic matter is present.

Transport in Soils

The transport of lead in soil is affected by (a) the specific adsorption or exchange at the particular mineral surface, (b) the formation of relatively stable complex ions or chelates with soil organic matter, and (c) the precipitation of sparingly soluble lead salts. Lead is deposited on soil by the dry and wet deposition processes, as previously discussed. Zimdahl and Arvik [78] have presented extensive data on cation exchange capacity, organic matter content, pH, soil type, and drainage as factors affecting the mobility of lead in soils. Lovering [44] has reviewed the interaction of organic chelating agents and their inconsistent ability to precipitate lead. Lead commonly reacts with soil anions (SO_4^{2-}, PO_4^{2-}, and CO_3^{2-}) or clays to form insoluble complexes, inhibiting mobility.

Physical, Chemical, and Photochemical Reactions

Physical Transformations

The environmental processes of transformation, transport, and deposition of lead are prominently influenced by the particle sizes. Size distributions of lead particulates have been studied to determine their relation to numerous evironmental questions [21, 27, 28, 34]. These investigations have determined that the smaller lead particles (less than 0.5 µm in diameter) make up the largest fraction of those exhausted from internal combustion engines, with mean particle size decreasing with increasing speed. There is little modification of the characteristics of the lead particulate size distribution as a result of non-precipitative atmospheric mechanisms.

Chemical Transformations

Huntzicker et al. [34] have shown the photochemical reactivity of tetraethyllead and tetramethyllead, finding that there is a light induced decomposition in the atmosphere. These organic compounds are less volatile than gasoline, but small amounts may be emitted to the atmosphere by evaporation from fuel storage and transportation systems.

Ter Haar and Bayard [70] have determined that the elevated temperature and pressure of combustion convert the lead alkyl compounds to lead oxides. The lead oxides subsequently react with the other fuel additives and leave the combustion chamber as a variety of complex compounds, with lead bromochloride (PbBrCl) being predominant. Pierrard [55] suggestes a photochemical decomposition of PbBrCl to form lead oxide and bromine and chlorine gas. The loss of halogen, but not the photochemical mechanism, is supported by other investigators [58, 70]. Lee et al. [42] has shown that the percentage of water-soluble particulate lead increased when exhaust was exposed to light in the near ultra-violet and visible regions.

Metabolism

Absorption

The absorption of lead from environmental sources depends on the amount of lead present, the physical state (usually characterized by particle size), and the chemical speciation. A host of factors influence the lead absorption in any individual, these include; age, physiological status, quantity of food eaten or air breathed, the proportionate ingestion or inhalation of lead, body weight, and the quantity of energy expended in day-to-day acitivity.

Inhalation

The International Radiological Protection Commission (IRPC) Task Group on Lung Dynamics [68] has developed an model designed to predict the percentage of inhaled aerosols that would be deposited and retained in the lungs. The model predicts that 35% of the lead inhaled in general ambient air would be deposited in the airways, based on particles with an aerodynamic diameter of 0.1–1.0 μm. Deposition of lead dusts generated from stationary (industrial) sources would be in the nasopharynx area, this is because these are usually larger particles. Lead aerosols breathed by the general population are not sufficiently characterized to predict deposition. This is particularly true for those particles less than 0.1 μm in diameter, which are deposited by diffusion [39].

Kehoe [36] studied the deposition of combusted tetraethyllead in human volunteers. Kehoe found 36% of particles with a mean diameter of 0.05 μm, and 46% of particles with a mean diameter of 0.9 μm were deposited in the lungs. These data were substantiated by Chamberlain et al. [9]. Under conditions of chronic airborne exposure approximately 50% of the deposited lead is absorbed. Beck et al. [4] have shown that although human alveolar macrophages ingest particles on the lungs, the cells may be damaged by inorganic lead compounds. Bingham et al. [5] has shown similar results in animal studies.

Gastrointestinal

The absorption of lead from food depends on the physical form of the dietary intake, with absorption from beverages being 5 to 8 times greater than from solid food [22]. Long term balance studies conducted by Kehoe [36] indicate that about 10% of the lead from food and beverages was absorbed through the gastrointestinal tract, since this was the amount excreted in the urine. This estimate disregarded intake by inhalation and did not measure lead in the feces.

Kostial et al. [37] demonstrated that 5–7 day old rats absorb at least 55% of single oral doses of Pb-203 while Forbes and Reina [20] reported that absorption was high prior to weaning but decreased rapidly thereafter. Garber and Wei [22] found that fasting enhanced lead absorption in mice. Low dietary levels of calcium, iron, zinc, copper, selenium, and vitamin D have been reported to enhance lead adsorption [64].

Cutaneous

In the case of organic compounds of lead, absorption through the skin is of some importance. Soon after it was marketed tetraethyllead was found to be easily absorbed through the skin, with the absorption process being reduced in the presence of gasoline [36]. If the skin is traumatized before application, absorption increases 3 to 4 times. Separation of cutaneous and respiratory exposure has been difficult.

Plants

In their study of two marine phytoplankton, Shultz-Baldes and Lewis [63] found the initial adsorption of lead onto the surface of plankton to follow the Freundlich adsorption isotherm. For the first hour of uptake, the lead ions were quickly and reversibly bound to cell surfaces, only later penetrating to deeper binding sites. After the first hour, and up to seven days, there was a reduction in extractable lead ions on phytoplankton. This was attributable to a decrease in the number of binding sites on protein, as a cell goes through its growth cycle.

In terrestrial plants the internal concentrations of lead are important in the limitation of primary production, but it is the total concentration that is of importance in toxicity to small herbivores and to animals higher in the food chain [40].

In the study of many terrestrial plants it was found that lead existed as a topical dust coating, fifty percent of which could be removed by a simple water washing [61]. The total lead concentration of each plant was found to increase with traffic volume and decrease with distance from the highway [49]. Because atmospheric lead is of great importance to total lead concentration in plants, surface to volume ratio is important in the topical lead coating of plants. Differences in internal lead concentrations in plants are due to the unique physiology of each species [16].

Lead occurring naturally in soil is the main source of lead found in the edible portions of many food crops, such as wheat, potatoes, tomatoes, sweet corn, carrots, cabbage, oats, rice, leaf lettuce, and snapp beans [69]. In all of these crops except leaf lettuce there is no measurable effect on the lead concentration in the edible portions due to atmospheric lead. The non-edible portions of these crops, such as corn husks, bean leaves, and the straw of grain crops showed up to a three-fold increase in lead concentration from atmospheric lead.

Distribution and Retention

A single dose of lead entering a body is distributed initially by the flow of body fluids to the various body compartments. There is then a redistribution of the lead to the various organs and systems, based on their respective affinities for lead. Over long periods of lead intake, a near steady state is achieved with respect to intercompartmental distribution [77].

Humans and Animals

The accumulation of lead begins in fetal life by transfer across the placenta, with an equilibrium between mother and child being attained. Autopsy data indicate that lead in humans becomes localized and accumulates in the bone, with the total body burden of men aged 60–70 years being 200 c.g. This total body burden is contained in two general pools. The bone is the major pool in terms of total lead, and is highly accumulative and has a relatively long exchange time. The second pool is comprised by the remaining organs and systems of the body. This second pool has a lower accumulation rate but a higher turnover rate than bone, reaching an equilibrium early in adult life. This second pool is also of greater toxicological significance, emphasizing the importance of this pool in the total body burden [73, 77].

A single dose of lead to rats initially produces high concentrations of lead in soft tissues, with a rapid reduction of this concentration as the result of transfer to the bone and excretion. The characteristics of the distribution of lead in animals has been found to be independent of the dose, over a wide range. As in humans there are two pools of lead, with lead being eliminated from bone much more slowly than from other tissues. Distribution in animals has been found to be dependent on age, species, length of time since dose application, and complex physiological variables. In rat studies lead was found to have an affinity for mitochondrial membranes but not for lysosomes, in contrast to mercury, copper, and iron [73, 77].

Plants

It has been shown that there is little translocation of absorbed lead in terrestrial plants, with lead being obtained both through the leaves and the roots [49]. Atmospheric lead was found not to accumulate in the edible portions of most plants, with strawberries and leaf lettuce being notable exceptions [61, 69]. The amount of lead absorbed through the roots of plants located near highways was found to be insignificant in the edible portions of plants.

The lead concentration of plants varies with the season of the year, because of variation in the growth rate of the plant [56]. When a constant influx of lead is maintained, the concentrations of lead found from April to June are lower because of the rapid growth. From June until autumn there is a slight increase in lead concentration, attributed to water loss, with undiminished growth. There is less growth in the winter months, while translocation continues, yielding an increase in lead concentration.

Plants are important in the biogeochemical cycle of lead [65]. Lead associated with leaves or other deciduous tissue is recycled relatively fast, while lead contained in woody parts of the plant is recycled over a much longer period of time.

Elimination

The elimination of lead from plants has not been studied. In animals the urine and feces are the main routes of elimination, with lesser quantities eliminated by sweat, hair growth, nail growth, and exfoliation of the skin.

The feces represents the major route of organic and inorganic lead elimination in man. The rate of fecal lead elimination is about 100 times the rate of elimination in urine, although most of the lead in feces represents metal that was not absorbed in the digestive tract. The characteristics of urinary lead excretion may be affected by the chemical form of the lead. Onyl one-third to two-thirds of the lead in the urine of lead workers is available for precipitation, indicating the presence of a stable lead complex. The lead concentration of sweat has also been found proportional to the dose. Although lead excretion appears to be disproportionally low in cases of high-level exposure, there is not yet a predictable relationship between increases in lead exposure and in lead excretion [73, 77].

The relative importance of lead transfer from the blood into the urine and feces, in animals, varies with the species tested. In sheep and rats, fecal elimination of lead is more rapid than urinary excretion [6], while urinary excretion is twice as great as fecal excretion in baboons [12]. In all of these cases, most of the lead excreted was derived from bile. Although species differences in gastrointestinal absorption rate and the presence of a gall bladder explain differences in the rate of appearance of a single dose of lead in the feces, these factors do not account for differences in the relative amounts of steady-state lead elimination in urine and feces [73].

Alkyl Lead Metabolism

Tetraethyl- and tetramethyl-lead do not themselves have toxic properties. The trialkyl compounds formed by dealkylation in the liver are the causes of toxic effects. A reduced toxicity of tetramethyl lead over tetraethyl lead is apparent because of the slower dealkylation rate of tetramethyl lead [73, 77].

Exposure and Accumulation

Occupation Exposures

The greatest potential for high-level exposure to lead exists in the process of lead smelting and refining [77]. In industry the major route of lead exposure is by inhalation of lead bearing dusts and the ingestion of objects on which these dusts have settled. The operations with the greatest hazard potential are those in which molten lead is vaporized, resulting in small respirable particles upon condensation. Thus the primary smelter blast furnace is a location where a great mass of respirable particles occurs. Lead mining hazards partially depend on the solubility of the ore. Thus galena (PbS), the major lead ore, being insoluble has minimal absorption in the lungs, although some may be converted to lead chloride in the stomach and absorbed. Proper ventilation is of utmost importance in reducing the potential of this hazard [77].

The heat produced in the welding or cutting of metals is sufficient to generate large quantities of respirable lead particulates from lead in the metal or from lead-containing coatings. Under poor ventilation, the electric arc welding of zinc silicate-coated steel has produced breathing concentrations in excess of the short term exposure limit set in the United States [77].

In the manufacturing of electric storage batteries, workers are exposed to high air lead concentrations (particularly lead oxide dust) in all stages except final assembly and finishing. Plate casting is a molten-metal operation in which there is spillage resulting in dusty floors. Frequent cleanup is necessary in the oxide mixing process to prevent dust accumulation [77].

Workers involved in the manufacture of tetraethyl- and tetramethyl-lead are exposed to both inorganic and alkyl lead. The major potential hazard, absorption through the skin, is reduced by the use of protective clothing [77].

Dietary Exposures

Most people receive the largest portion of their daily lead intake through foods, with estimates of the daily dietary intake of lead in males ranging from 100 to 500 mg/day [60], but only a fraction of this amount is absorbed by the body. As discussed earlier the concentration of lead in crops depends on a number of factors

including traffic density and soil conditions [49]. The amount of lead a person in-
gests with food depends on (a) the total amount of food eaten, (b) the growth his-
tory of the food, (c) the foods opportunity to absorb lead from water, soil, or topical deposition, and (d) personal dietary habits (e.g. a preference for fresh vegetables) [73].

On a per-weight basis, children have two to three times the dietary lead intake
of adults. This high value for children is attributed to the high consumption of food
stored in cans. The soldered seam of tin cans is evidently the major source of added
lead in canned foods, with lead concentrations of food in the cans increasing as
samples are taken closer to the seam [48]. There is also an increasing total lead concentration in a can with the increasing ratio of the cans seam length to volume. The
colored portions of wrappers from bakery confections, candies, and gums are also
potential sources of lead exposure [30]. Harris and Elsea [31] have shown that lead
poisoning can result from the storage of an acidic beverage in improperly glazed
earthenware vessels.

The United States Public Health Services drinking water standards specify a
50 mg/l maximum for lead, EPA [73] has shown that this limit is only infrequently
exceeded in larger United States cities. The presence of lead in drinking water may
result from the use of lead materials in the water distribution system. Naturally occurring lead in rocks and soils may be an important source of contamination in isolated instances but lead from industrial wastes represent a local and not a widespread problem. The disposition of lead compounds from gasoline is a major
source of lead in water systems [73].

Ambient Air Exposures

The levels of atmospheric lead are gradually decreasing, mainly as a result of the
decreased use of lead in gasoline. As is expected, urban atmospheres contain higher
lead concentrations than rural atmospheres. In 1974, the arithmetic mean of urban
lead concentrations was $0.89 \mu g/m^3$, as compared to the rural mean of $0.11 \mu g/m^3$.
When rural locations are classified according to their proximity to large population
centers, the lead concentrations decrease with distance from the urban environment [73].

Accumulation of Lead in Humans

The concentration of lead in blood is used as an index of exposure to assess conditions considered to represent a risk to health. Plasma lead concentrations have
been shown to be constant at $0.2–0.3 \mu g/l$ over a range of $100–1,500 \mu g/l$ in whole
blood showing that lead is bound primarily to erythrocyte protein (chiefly hemoglobin) rather than to stroma [73].

Lead concentrations in teeth and hair have been used as indicators of integrated
long-term exposure, having the advantage of easy sample procurement. As yet, the
amount of information concerning the interpretation of these data is inadequate
for their evaluation as indices of exposure or dose [77].

Relationships Between External Exposures and Blood Lead Levels

Studies indicate that the blood lead to air lead ratio is not constant over the range of concentrations normally encountered. Most of the studies indicated blood to have one to two times the concentration of air, with males having a slightly higher value than females, and children a slightly higher value than adults. More data are needed in the study of blood-air ratios in low air lead concentrations [73].

Observable increases in blood lead levels occur at soil or dust lead concentrations of 500–1,000 mg/l. There is a consistent 3%–6% increase in blood lead levels for a two-fold increase in soil lead levels, over a wide diversity of populations. In data observed, the largest estimate in the percentage increase in blood lead occuring in populations with the lowest lead levels [73].

In the analysis of two studies, for 40%–45% of confirmed cases of elevated blood lead levels in children, a possible source of lead paint hazard could not be located [73]. However, that should not weaken the role of lead-based paint as a major source of lead for children. The background contribution of lead from other sources is still not known even for children whose potential lead-paint hazard has been identified, making a discrimination of the lead proportion from either source impossible.

Persistence

The residence time of lead compounds predicts their accumulation and redistribution in the atmosphere, hydrosphere, and lithosphere. McDonald [47] has estimated the average residence time for aerosols in urban-industrial areas of the USA to be 10 h, with some locations as low as 2 h. For urban, non-industrial, areas a range of 3–26 h was determined. Both sets of data assumed mixing heights of 100–1,000 m. These data are supported by Edgington and Robbins [19] who estimate a 10 h residence time above Lake Michigan, and Winchester et al. [75] who found the residence time of particulate lead halide to be less than one day.

Goldberg and Arrhenius [24] determined that 2.5×10^{15} g of weathered products and pyroclasts enter the ocean each year. A residence time for lead was calculated to be 2,000 years, assuming the lead content in sediments to be the same as in crustal rocks. Durum and Haffty [18] calculated a residence time of 600 years for lead entering the ocean from rivers.

Biological Effects and Toxicity

Effects and Toxicity to Man

Epidemiology of Lead Poisoning at the Population Level

In a seven year study of school aged children in the United States, Angle and McIntire [2] concluded that dust and soil fallout of airborne lead as well as dietary and unusual sources are all important in determining body burden. Intestinal absorption of lead was found to be as high as 42%, and in the absence of unusual sources of lead, is the principal component of the body burden. Community-wide changes in blood lead were found to be multifactorial, with air, soil, water, housing, and socio-economic shifts having an additive or even a synergistic effect.

In a study of 62 workers in a battery factory, Richter et al. [57] found that blood lead levels of 550–650 µg/l caused medical and subclinical problems, with biochemical and insidious neuro-physiologic changes occurring at lower blood lead levels. They recommended that occupational 8-h exposures be lowered to 50 µg/m³ to keep blood lead levels in safe ranges.

Toxicity

Haemopoietic System. Lead poisoning causes anemia by impairment of heme synthesis and by an increased destruction rate of redblood cells. Some of the steps in heme syntheses occur in the mitochondria, a site of lead accumulation. It is in the mitochondria that lead impairs the incorporation of iron molecules into the heme. Lead also decreases the life span of red blood cells but the mechanism responsible for this effect is not well understood [25].

Nervous System. Structural and functional nervous system effects of lead are apparent. These effects involve the brain, cerebellum, spinal cord, and motor and sensory nerves leading to specific areas of the body. Neural tissue is very sensitive to the toxic effects of lead with cases of brain swelling reported at even low lead concentrations. Severe subcellular damage is usually noted in the cerebellar cortex and cerebral cortex, and sometimes in the basal ganglia. This damage also impairs the flow of blood in the brain [25].

Renal System and Gastrointestinal Tract. The kidney is an excretory organ and therefore has a prominant role in lead metabolism. Lead has been found to cause a reduction in kidney transport of amino acids, glucose, uric acid, citric acid, and phosphate; probably because of a decrease in energy production. This decrease in energy production is caused by a direct effect of lead on renal mitochondria, and enzymes responsible for energy production [25].

An early symptom of lead poisoning is colic, warning that more serious effects may occur with continued and prolonged exposure. Colic is commonly noticed in industrial exposure cases and in lead poisoning of infants and young children. Although there is extensive documentation in the literature, data are insufficient to establish a dose-response relationship for the effect of lead on the gastrointestinal tract [73].

Cardiovascular System. Dingwall-Fordyce and Lane [17] found marked increase in cerebrovascular mortality among heavily exposed workers. These exposures were in the first quarter of this century, a time of poor working conditions. A similar increase was not found in the mortality of unemployed males. Other studies have concluded that there was excess mortality associated with only two illnesses, chronic nephritis and hypertension. It is unclear whether vascular effects of lead in man are direct effects on blood vessels or if the effects are secondary to renal effects [73].

Reproductive System. Severe exposure to lead has been associated with sterility, abortion, still births, and neonatal deaths [25], although there are no epidemiologic data [77]. Women working in lead industries were found to have a higher rate of ovulatory dysfunction, and men with moderately increased lead absorption were found to have decreased fertility.

Endocrine Organs. Although lead is known to decrease thyroid function in man, the endocrine effects of lead are not well-defined. Lead may also interfere with pituitary function in man, producing hypopituitarism in some individuals. Excessive oral ingestion may cause pathological changes in the pituitary-adrenal axis, decreasing metapyrone response and pituitary reserve [59, 73].

Carcinogenicity

Benign and malignant renal neoplasms have been observed in rats and mice fed diets with 100 or 1,000 mg of basic lead acetate per kg of diet. The results of these studies have been summarized [77]. Tumors in the testes and in the adrenal, thyroid, pituitary, and prostate glands have been noted in addition to renal neoplasms. Intraperitoneal and subcutaneous injections of lead phosphate have also caused the development of renal tumors in rats. Tetraethyl lead has been shown to produce malignant lymphomas in mice [77]. Dingwall-Fordyce and Lane [17] did not find any increase of malignant diseases in their follow-up study of 267 workers.

Mutagenity

Conclusions vary regarding chromosome abnormalities resulting from exposure to lead. Chromosomal aberrations have been reported to result from lead exposure causing blood lead concentrations of 380–750 µg/l [77], while O'Riordan and Evans [52] did not find any significant increase in chromosomal aberrations in shipbreakers with blood lead levels of 400–1,200 µg/l. The occurrence of chromosomal abnormalities from lead exposure is therefore uncertain.

Factors Influencing Lead Toxicity

There appears to be no acquisition of tolerance to continuous exposure to lead. Measurement of metabolic parameters indicate that a stabilization occurs, but no return to normal values is noted [77]. Young people absorb lead faster and are more susceptible to its toxic effects than adults. However, the susceptibility of the elderly compared to younger adults has not been studied. People with hemoglobin and erythrocyte anomalies or renal damage are probably more sensitive to the effects of lead exposure. It has also been shown that heavy drinkers among industrially exposed men are more prone to lead toxicity [77].

Effects and Toxicity to Plants

Inside of plant cells, lead can have effects on the two major processes of a plant, i.e. photosynthesis and metabolism. Many of these effects are the result of lead forming as a coating on plant surfaces, causing an interference with gas exchange capacities of the plants [3].

In their study of corn and soybeans, Bazzaz et al. [3] found a decrease in net photosynthesis and transpiration with increasing lead treatment levels from $PbCl_2$. At lower lead concentrations corn, a C_4 plant, was more sensitive. At higher lead concentrations soybeans, a C_3 plant, was more sensitive, and was found to have lower photosynthesis and transpiration rates. The rate changes of the two processes are related to changes in the leaf stomatal resistance to CO_2, and water vapor diffusion [3]. It is believed that the difference in storage areas of photosynthesis products in C_3 and C_4 plants causes this difference.

Another study of soybeans indicated that when plants were grown in media with an increasing lead concentration, respiration rates increased up to 60 mg/l of lead in the culture media, with rates decreasing above this concentration [41]. Plants treated with lead had a reduced photosynthetic rate and a reduced photosynthetic phosphorylation rate.

Activity of acid phosphatase and peroxidase both increase with increasing lead concentration (with data up to 100 mg/l concentration in the culture media) [41]. These increased activities reflect the onset of a senescence response. Peroxidase is a heme-carrying enzyme; the increase in its activity with increasing concentrations of lead in the culture media indicates that lead levels up to 100 mg/l do not affect iron metabolism in soybean leaves.

With increasing lead concentration there is a shift from normal synthetic activity to a degradation process [41]. Although total free amino acids are unaffected by lead concentration, there is an increase in soluble protein with increasing lead concentration, which corresponds to the increase in activity of the hydrolitic enzymes.

Acknowledgement

The authors would like to thank the Applied Ecology Department, Operations Analysis Division, and Professional Development Award Committee of the Research Triangle Institute and the Texas Christian University Research Foundation for their support in preparation of the manuscript. We would especially like to thank Carol Nikodem and Hall Ashmore for their assistance.

References

1. Alder, H.H., Kerr, P.F.: Am. Min. *48*, 124 (1963)
2. Angle, C.R., McIntire, M.S.: J. Toxic. Envir. Health *5*, 885 (1979)
3. Bazzaz, F.A., Rolfe, G.L., Windle, P.: J. Environ. Qual. *3(2)*, 156 (1974)
4. Beck, E.G., Manojilovic, N., Fisher, A.B.: Die zytotoxiziat von blen. In: Proc. Internat. Symposium, Environm. Health Aspects of Lead, Amsterdam/Luxembourg, Commission of the European Communities, 1973, p. 451

5. Bingham, E. et al.: Science *162*, 1297 (1968)
6. Blaxter, K.L., Cowic, A.T.: Nature (Lond.) *147*, 588 (1946)
7. Bowen, H.J.M.: Trace Elements in Biochemistry. New York, Academic Press 1966
8. Bowen, H.J.M.: Environmental chemistry of the Elements. New York, Academic Press 1979
9. Chamberlain, A. et al.: Proc. Roy. Soc. London *192*, 77 (1975)
10. Cholak, J.L., Schafer, L.S., Yeager, D.: J. Am. Ind. Hyg. Assoc. *29(6)*, 562 (1968)
11. Chow, T.J.: Lead in natural waters. In: The Biogeochemistry of Lead in the Environment, Part A. (Nriagu, J.O., ed.), Amsterdam-New York-Oxford, Elsevier/North-Holland Biomedical Press 1978, p. 185
12. Cikrt, M.: Brit. J. Ind. Med. *29*, 74 (1972)
13. Cotton, F.A., Wilkinson, G.: Adv. Inorganic Chem., New York, John Wiley and Sons Inc. 1972
14. Daines, R.H., Motto, H., Chilko, D.M.: Envir. Sci. Technol. *4(4)*, 318 (1970)
15. Davis, W.E.: Emission study of industrial sources of lead air pollutants, US EPA, Document APTD-1543, 1973, p. 1
16. Dedolph, R.G. et al.: Environ. Sci. Technol., *4(4)*, 217 (1970)
17. Dingwall-Fordayce, J., Lane, R.E.: Brit. J. Ind. Med. *20*, 313 (1963)
18. Duram, W.H., Haffty, J.: Geochim. Cosmochim. Acta *27*, 1 (1963)
19. Edgington, D.N., Robbins, J.A.: Envir. Sci. Technol. *10(3)*, 226 (1976)
20. Forbes, G.B., Reina, J.C.: Proc. Soc. Exp. Biol. Med., *142*, 471 (1972)
21. Ganley, J.T., Springer, G.S.: Envir. Sc. Technol. *8(4)*, 340 (1974)
22. Garber, B.T., Wei, E.: Toxic. Appl. Pharm. *27*, 685 (1974)
23. Garrels, R.M., Christ, C.L.L.: Solutions, Minerals, and Equilibria, New York, Harper and Row 1965
24. Goldberg, E.D., Arrhenius, G.O.: Geochim. Cosmochim. Acta *13*, 153 (1958)
25. Goyer, R.A., Chisolm, J.J.: Lead. In: Metallic Contamination and Human Health (Lee, D.H.K., ed.). New York and London, Academic Press 1972, p. 57
26. Grandjean, P., Nielsen, T.: Residue Rev. *72*, 98 (1979)
27. Habibi, K.: Envir. Sci. Technol. *4(3)*, 239 (1970)
28. Habibi, K.: Envir. Sci. Techn. *7(3)*, 223 (1973)
29. Handbook of Chemistry and Physics, 55th Ed., CRC Press, 1974
30. Hankin, L., Heichel, G.H., Botsford, R.A.: Clin. Pediatr., *12*, 654 (1973)
31. Harris, R.W., Elsea, W.R.: J. Am. Med. Assoc. *202*, 544 (1967)
32. Hatch, W.R., Ott, W.L.: Anal. Chem. *40*, 2085 (1965)
33. Hilburn, M.E.: Chem. Soc. Rev. *8(1)*, 63 (1979)
34. Huntzicker, J.J., Friedlander, S.K., Davidson, C.I.: Envir. Sci. Technol. *9(5)*, 448 (1975)
35. Johansson, T.B. et al.: Elemental trace analysis of small samples by proton-induced X-ray emission. Anal. Chem. *47(6)*, 855 (1975)
36. Kehoe, R.A.: The metabolism of lead in health and disease. The Harben Lectures. J. Roy. Inst. Publ. Health Hyg., *24*, 81, 101, 129, 177 (1961)
37. Kostial, K., Simonovic, I., Pisonic, U.: Nature (Lond.), *233*, 564 (1971)
38. Langheinrich, A.P., Roberts, D.B.: Optical emission spectroscopy, In: Modern Methods of Geochemical Analysis (Wainerdi, R.E., Ulkew, E.A., eds.). New York-London, Plenum Press 1971, p. 769
39. Lawther, P.J. et al.: Airborne lead and its uptake by inhalation. In: Lead in the Environment (Hepple, P., ed.), Essex, UK, Applied Science Publishers 1972, p. 8
40. Lee, J.A.: Nature (Lond.) *328*, 165 (1972)
41. Lee, K.C. et al.: J. Environ. Quality *5*, 357 (1976)
42. Lee, R.E. et al.: Atmos. Environ. *5(4)*, 225 (1971)
43. Liebhafsky, H.A., Pfeiffer, H.G.: X-ray techniques. In: Modern Methods of Geochemical Analysis. (Wainerdi, R.E., Uken, E.A., eds.). New York-London, Plenum Press 1971, p. 245
44. Lovering, G.T.: Lead in the Environment. USGS: Washington, D.C. Professional Paper No. 957, 1976
45. Lutz, G.A. et al.: Technical, Intelligence, and Project Information System for the Environmental Health Service, Vol. III. Lead Model Case Study. National Technical Information Service: Springfield, Va. 1970
46. Manahan, S.E.: Environmental Chemistry, 2nd Ed. Boston, Willard Grant Press 1975
47. McDonald, J.E.: Bull. Am. Meteoral. Soc. *42*, 664 (1961)

48. Mitchell, D.G., Aldous, K.M.: Environ. Health Perspect. 7, 59 (1974)
49. Motto, H.L. et al.: Envir. Sci. Technol. 4(3), 231 (1970)
50. Nriagu, J.O.: Lead in soils, sediments and major rock types. In: The Biogeochemistry of Lead in the Environment, Part A. (Nriagu, J.O., ed.). Amsterdam-New York-Oxford, Elsevier/North-Holland Biomedical Press 1978, p. 15
51. Olson, K.W., Skogerboe, R.K.: Envir. Sci. Technol. 9(3), 227 (1975)
52. O'Riordan, M.L., Evans, H.J.: Nature (Lond.) 247, 50 (1974)
53. Parker, G.A., Boltz, D.F.: Colorimetry. In: Modern Methods of Geochemical Analysis (Wainerdi, R.E., Uken, E.A., eds.), New York-London, Plenum Press 1971, p. 97
54. Pasquill, F.: Atmospheric Diffusion, 2nd ed. New York, Halsted Press 1974
55. Pierrard, J.M.: Envir. Sci. Technol. 3(1), 48 (1969)
56. Rains, D.W.: Nature 233, 210 (1971)
57. Richter, E.D., Yaffee, Y., Gruener, N.: Envir. Res. 20, 87 (1979)
58. Robbins, J.A., Snitz, F.L.: Envir. Sci. Technol. 6(2), 164 (1972)
59. Sandstead, H.H. et al.: Clin. Res. 18, 76 (1970
60. Schroeder, H.A., Tipton, I.H.: Arch Environ. Health 17, 965 (1968)
61. Schuck, E.A., Locke, J.K.: Envir. Sci. Technol. 4(4), 318 (1970)
62. Shapiro, H., Frey, F.W.: The Organic Compounds of Lead, New York, Interscience 1968
63. Shultz-Baldes, M., Lewis, R.A.: Bio. Bull. 150, 118 (1976)
64. Six, K.M., Goyer, R.A.: J. Lab. Clin. Med. 76, 933 (1970)
65. Smith, William, H.: Science 176, 1237 (1972)
66. Stumm, W., Bilinski, H.: Trace metals in natural waters: Difficulties of interpretation arising from our ignorance on their speciation. In: Advances on Water Pollution Research (Jenkins, S.H., ed.). Proc. 6th Int. Conf. Jerusalem, New York, Pergamon Press 1972, p. 32
67. Stumm, W., Morgan, J.J.: Aquatic chemistry. New York-London-Sydney-Toronto, Wiley-Interscience 1970
68. Task Group on Lung Dynamics: Health Phys. 12, 173 (1966)
69. Ter Haar, G.: Envir. Sci. Techn. 4(3), 226 (1970)
70. Ter Haar, G.L., Bayard, M.A.: Nature (Lond.) 232, 553 (1971)
71. Ter Haar, G.L., Holtzman, R.B., Lucas, H.F.: Nature (Lond.) 216, 353 (1967)
72. Tuddenham, W.M., Stephens, J.D.: Infrared spectrometry. In: Modern Methods of geochemical Analysis. (Wainerdi, R.E., Uken, E.A., eds). New York-London, Plenum Press 1971, p. 127
73. US Environmental Protection Agency: Air Quality Criteria for Lead. USA 1977
74. Willard, H.H., Merritt, L.L., Dean, J.A.: Instrumental Methods of Analysis, 5th ed. New York, D. VanNostrand 1974
75. Winchester, J.W. et al.: Atmosph. Environ. 1, 105 (1967)
76. Wixson, B.G., Jennett, J.C.: Envir. Sci. Tech. 9(13), 1128 (1975)
77. World Health Organization: Environmental health criteria, p. 3. Lead. Geneva, Switzerland 1977
78. Zimdahl, R.L., Arvik, J.H.: Crit. Rev. Environ. Control 3, 213 (1973)

Arsenic, Beryllium, Selenium and Vanadium

L. W. Newland

Environmental Sciences Program, Texas Christian University
Fort Worth, TX 76129, USA

Introduction

The importance of toxic elements in environmental chemistry is rarely questioned, but a relatively small number of elements (mercury, lead, and cadmium) have received a large share of researchers' attention. The environmental chemistry of the transition metals, e.g., chromium, nickel, manganese, cobalt, copper, etc., has also been investigated principally because of their roles in metabolism, especially enzymatic processes. However, two non-metals, arsenic and selenium, and two metals, beryllium and vanadium, are elements which will become more significant in the future from environmental and toxicological points of view. Arsenic and selenium have been investigated, but much more work is needed because of the importance of these two elements in the environment. The author considers beryllium and vanadium to be "problem metals of the future". The primary exposure route for both beryllium and vanadium is via the atmosphere and as lower environmental standards are imposed, more uses are found for each element, and more fossil fuels (source of V) are burned, the amounts added to the atmosphere will have more significance.

This chapter is an attempt to assemble the most salient information concerning arsenic, beryllium, selenium, and vanadium and bring these elements to the attention of environmental scientists. The chapter is divided into four sections, one for each element, and follows similar topic outlines wherever possible.

Arsenic

The alchemist's symbol for arsenic, a coiled serpent, symbolizes very well the element's prevailing evil reputation. Arsenical compounds were used in medicine in the Orient 2,000–3,000 years ago. The Greek philosopher, Theophrastus (c. 370–287 BC), wrote of it, though discovery of the element is generally credited to Albertus Magnus (c. 1206–1280). Arsenic has played a large role in human medicinal and evil practices for many years. Arsenic was the preferred homicidal and suicidal agent during the middle ages.

Production, Use and Shipment

Arsenic is present in all copper, lead, and zinc sulfide ores and is carried along with those metals in the mining, milling, and concentration processes. Arsenic trioxide (As_2O_3), which is the arsenic of commerce, is a basic raw material for herbicides, fungicides, insecticides, algicides, sheep dips, wood preservatives, feed additives, and human and veterinarian medicinals [11]. Elemental arsenic is useful in alloys, particularly lead alloys for shot or lead battery grids, and it is a constituent element of gallium arsenide, which is responsible for the colors of digital watches. It is also used in the light-emitting diodes of other instruments.

The world production of arsenic trioxide was about 50,000 tons in 1977. The US comsumes about half the total world production, and produces about 50% of what it consumes [11]. According to the National Academy of Sciences, exporters of arsenic include Mexico, Sweden, France, the Republic of South Africa, Peru, and the Phillipines [10]. The amounts of arsenic in the various copper, lead, and zinc concentrates range from parts per million (ppm) up to 15.5%.

Agricultural uses account for most of the world consumption of arsenic. Arsenic trioxide is the raw material for arsenical pesticides, including lead arsenate, calcium arsenate, sodium arsenite, and organic arsenicals. Compounds like Paris green (copper acetoarsenite) were formerly popular insecticides in orchards, but are only of minor importance today. Lead arsenate and calcium arsenate have been used extensively for insect control on fruits, tobacco, cotton, and some vegetables, but current use is slight [10].

Sodium arsenite came into use as an insecticide between 1920 and 1930, mainly as a bait and livestock dip. As an insecticide it is highly phytotoxic. As a cattle dip it is used to control ticks, fleas, and lice. The inorganic arsenicals have been in use since 1890 as weedkillers, particularly as non-selective soil sterilants. Today these compounds have been replaced by organic arsenicals. Arsenic acid is used extensively in cotton-growing areas as a dessicant to facilitate mechanical harvesting [13]. Dessicants are used to prepare cotton plants for stripper harvesting by depleting the leaves and other plant parts of moisture, thereby improving harvesting efficiency and preventing degradation of fiber quality that results from leaf staining.

Organic arsenicals such as arsenilic acid, 3-nitro-4-hydroxyphenylarsonic acid, 4-nitrophenylarsonic acid, and p-ureidobenzenearsonic acid have been used as a feed additive since the 1950's [4]. The U.S. Food and Administration approves of all four of these compounds for use in poultry and swine feeds but only at levels low enough to preclude residues in edible animal tissues which would be hazardous to human health [10]. In France and other countries almost all uses of these compounds are forbidden.

Chemistry

Arsenic has an atomic weight of 74.9216, atomic number 33, and valences of 5, 3, 0, −3. Elemental arsenic occurs in two solid variations: yellow, and gray or metallic, with specific gravities of 1.97 and 5.73, respectively. Gray arsenic, the ordinary stable form, has a melting point of 817 °C (at 28 atm), and sublimes at 613 °C [18]. Arsenic is found native, in the sulfides realgar and orpiment, as arsenates. Mispickel or arsenopyrite (FeSAs) is the most common mineral, from which of heating, the arsenic sublimes leaving ferrous sulfide. The element is a steel gray, very brittle, crystalline, semimetallic solid. It tarnishes in air, and when heated is rapidly oxidized to arsenous oxide (As_2O_3), with the odor of garlic.

Analytical Methods

Analytical methods for the determination of arsenic are numerous. Some of the more important methods include atomic absorption spectrophotometry, molecular-absorption spectrophotometry, atomic-emission spectroscopy, gas chromatography, and co-precipitation. It is now possible to determine several species of arsenic in natural waters using a method based on the sequential volatilization technique where the species are volatilized as arsines [2]. The arsines are then separated either by fractional volatilization or by gas chromatography, and detected either by atomic absorption, electron capture, or flame ionization. The detection limit for AsH_3 using graphite furnace and atomic absorption is about 0.05 ng As and the detection limit for dimethyl arsine using an electron capture detector is 0.2 ng As. Using the flame ionization detector for dimethyl arsine, one can achieve a detection limit of 1.0 ng As. Using neutron activation analysis, arsenic can be simultaneously determined with antimony, cadmium, chromium, copper, and selenium in environmental material [7]. One of the advantages of neutron activation is that it is non-destructive to the sample. Arsenic can also be quantitatively analyzed with mercury and bromine in atmospheric aerosols using neutron activation [9].

Transport Behavior

The principal pathways that arsenic follows from the continents to the oceans in the absence of human interference are weathering, including solubilization and transport of sediment, and vulcanism. Human activities greatly influence the amount of arsenic to the environment. Indirect releasing of arsenic comes from burning of fossil fuels, manufacturing of arsenicals, and erosion of the land. Table 1 summarizes the molecular forms of arsenic as found in the environment and Fig. 1 represents the environmental chemistry of arsenic.

Sandberg and Allen [14] proposed a model (Fig. 2) for arsenic cycle in an agronomic ecosystem. This model contains 12 possible transfers to and from a field for the organoarsenical herbicides. They conclude that transfers involving reduction to methylarsines, soil erosion, and crop uptake were primary redistribution mechanisms in this model. From their data it was concluded that arsenic is mobile and nonaccumulative in the air, plant and water phases of the agronomic ecosystem. Arsenicals do accumulate in soil but redistribution mechanisms preclude hazardous accumulations at a given site.

High concentrations of arsenic (maximum 2,100 ppm; average 115 ppm; median 60 ppm) have been found in sediments from the areas of hot brines in the Red

Table 1. Environmental forms of arsenic [17]

Compound	Source
Water	
As(III), arsenite ion and	Sea water
As(V), arsenate ion	Fresh water ponds, rivers, lakes
$CH_3AsO(OH)_2$	Sea water, fresh water ponds, rivers, lakes
$(CH_3)_2AsO(OH)$	Sea water, fresh water
$(CH_3)_3As$ (or the oxide)	Fresh water
Air	
As(III) and As(V)	Particulate
CH_3AsH_2	Over As-treated soil
$(CH_3)_2AsH$	Over-treated soil
$(CH_3)_3As$	Over-treated soil
Biological samples	
Type	Forms
Sea weed and epiphytes	As(III), As(V), $CH_3As(OH)_2$, $(CH_3)_2AsO(OH)$, $(CH_3)_3As$
Urine	As(III), As(V), $CH_3AsO(OH)_2$, $(CH_3)_2AsO(OH)$
Methanobacterium cultures	$(CH_3)_2AsH$
Aerobic cultures (Fungi and mixed)	$(CH_3)_3As$, $CH_3AsO(OH)_2$, $(CH_3)_2AsO(OH)$

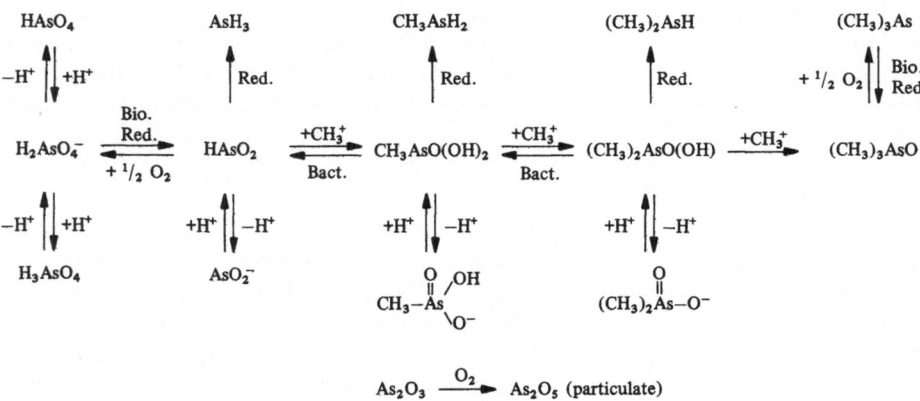

Fig. 1. Summary of environmental chemistry of arsenic [17]

Sea. Randama and Sahama [12] suggest that sediments and sedimentary rocks of marine origin should contain more arsenic than igneous rocks do, because large amounts of arsenic have been introduced directly into the exogenic cycle by volcanic activity.

Inputs into the environment and a redistribution of arsenic in the terrestrial ecosystem are presented in Fig. 3. Natural inputs are from decay of plant matter, volcanic action, and weathering of minerals within the soil, whereas man-made sources of arsenic are combustion of coal and oil, smelting of ores, and use of fer-

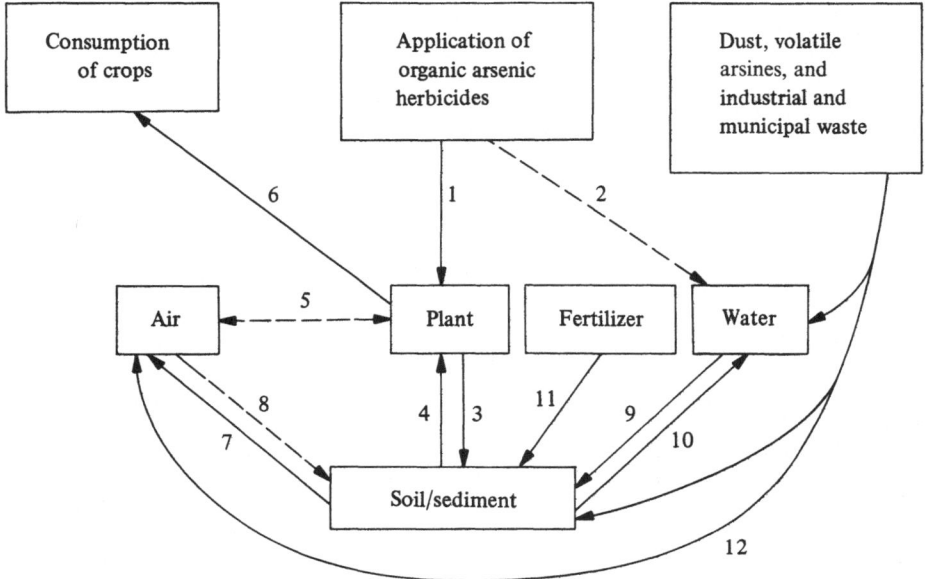

Fig. 2. A proposed model for the arsenic cycle in an agronomic ecosystem [14]

tilizers and pesticides [10]. The largest sink for arsenic from human activity in the environment is the soil.

As a result of soil erosion, industrial and agricultural application, arsenic and it's compounds are widely distributed through the aquatic environment [16]. Arsenic can be removed from industrial waste by several methods before the waste is discharged into the water system. Some of the methods include precipitation by calcium oxide and ferric chloride, basic anion exchange resins, passage through lime and ashes, and flocculation with chlorine saturated water and ferrous sulfate [10].

Table 2 calculates the amount of arsenic emission to the atmosphere from various sources of arsenic [10].

Carton, as cited by [10] surveyed arsenic input and movement in the US. He estimated a total movement of about 119,000 tons of arsenic per year (Table 3). He distinguished between arsenic found as end products and arsenic that is dissipated onto land, emitted in air and water, or destined for landfills. Of the 119,000 tons, most is fixed in products in which the arsenic is immobile or is deposited in land fills as waste material. The remainder is in a form that can move readily within the environment.

A cycle for arsenic in a stratified lake is shown in Fig. 4. The reactions include transfers from solution to solid phases, conversions from one oxidation state to another, and ligand exchanges. Some of the processes are chemical, some occur through microbial mediation, and some can occur either way. Fish and plants enter the cycle by concentrating arsenic, especially trimethylarsine. Upon death the organisms settle to the bottom where the arsenic is removed to the sediments or recycled depending on the physical and chemical conditions [6].

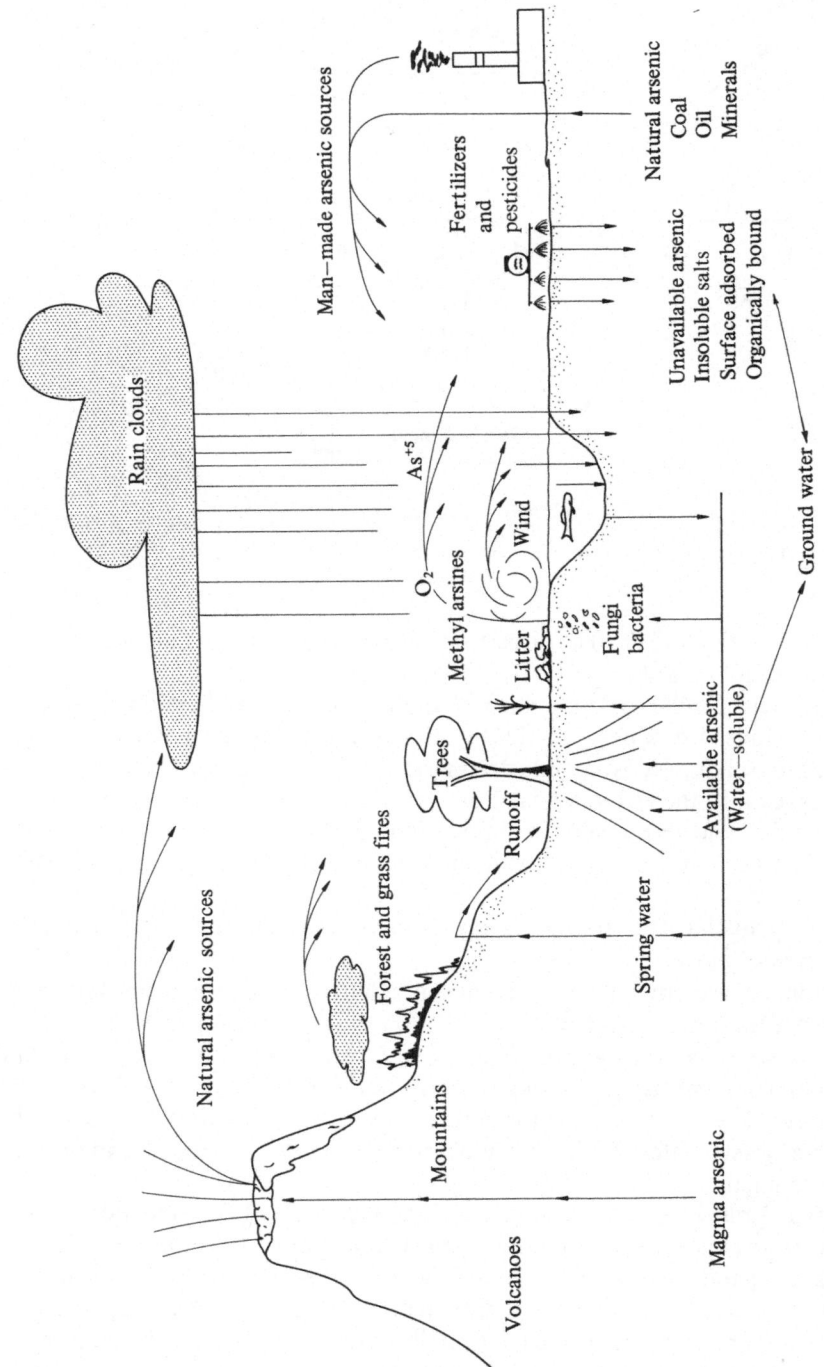

Fig. 3. Environmental transfer of arsenic [10]

Table 2. Arsenic emission factors for the United States (1968) [10]

Arsenic source	Arsenic concentration
Mining and milling	0.45 tons/million tons Cu, Pb, Zn, Ag, Au, or U ore
Smelting and refining	955 tons/million tons of Cu produced
	591 tons/million tons of Zn produced
	364 tons/million tons of Pb produced
Coal	1.4 tons/million tons coal burned
Petroleum	5.2 kg/million barrels of petroleum

Table 3. Summary of U.S. arsenic flow, dissipation, and emission (1974)

Location of arsenic	Arsenic flow, tons	Ready environmental transport
End products	26,438	
Steel	17,089	No
Cast Iron	3,368	No
Other	5,711	No
Dissipation to land:	63,030	
Steel slag	39,690	Unknown
Pesticides	11,565	Yes
Copper leach liquor	9,702	Yes
Other	2,073	Yes
Airborne emission:	9,757	
Losses from copper-smelting	5,292	Yes
Pesticides	2,536	Yes
Coal	717	Yes
Other	1,212	Yes
Waterborne effluent:	165	
Phosphate detergents	121	Yes
Other	44	Yes
Landfill wastes:	19,691	
Copper flue dusts	10,584	No
Copper-smelting slag	3,748	No
Coal fly ash	1,984	No
Other	3,375	No

It can be concluded that arsenic is continuously cycling in the environment through oxidation, reduction, and methylation reactions. Man's activities can alter the distribution of arsenic in specific geographic areas or in selected components of the environment, but has little or no control over the natural processes. Arsenic in the aquatic systems has an unusually complex and interesting chemistry with oxidation-reduction, ligand exchange, precipitation, and adsorption reactions all taking place [6]. Arsenic is stable in all four oxidation states under Eh conditions occurring in aquatic systems. Arsenic metal occurs only rarely, and − 3 arsenic only at extremely low Eh values. The rate of oxidation of arsenite to arsenate with oxygen is reported to be very slow at neutral pH values: but the reaction proceeds

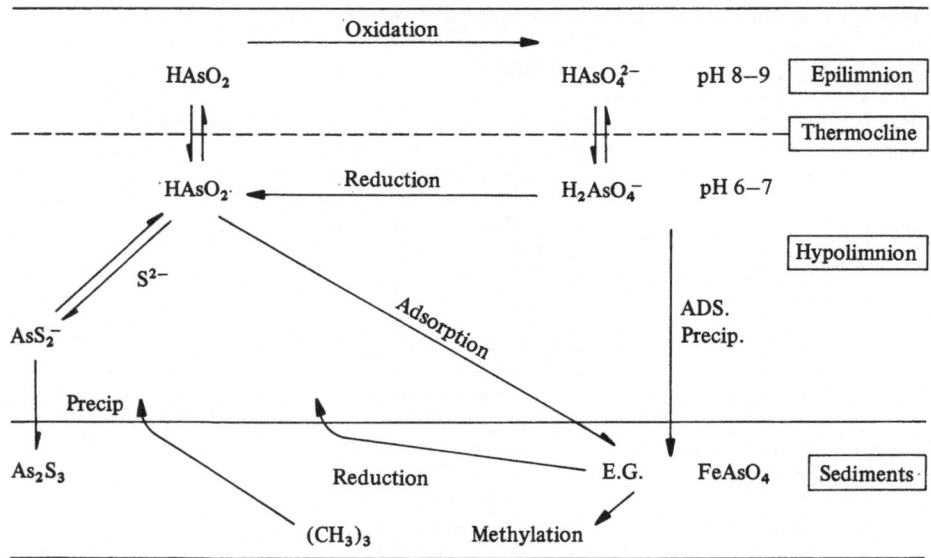

Fig. 4. Local cycle of arsenic in a stratified lake (from [6])

measurably in several days in strongly alkaline or acid solutions. Arsenic $+3$ and $+5$ both form compounds with carbonarsenic bonds. Thousands of these compounds have been synthesized and tested for effectiveness against various pests, weeds, insects, and human parasites [6]. Some of these compounds persist in water, others are rapidly oxidized, hydrolyzed or otherwise degraded.

Extensive studies have shown that arsenic can accumulate in the soil when arsenic compounds are repeatedly applied to crops [1]. On fields treated with calcium arsenate for insect control, the arsenic concentrations decreased with soil depth [1]. Appreciable amounts of arsenic could also move down in the soil profile with the percolating soil water [13].

Accumulation and Metabolism

Large soil accumulations of arsenic occur around smelters. In a study done in S.W. England, the grossly contaminated sites are largely barren and support only a restricted number of species [19].

Arsenicals will accumulate in soil when greater amounts are added than removed in harvested plants, through volatilization and through leaching. As mentioned before arsenic and its compounds can be accumulated by organisms in the aquatic environment, but this accumulation generally does not occur in the food chain.

Aquatic organisms are able to accumulate arsenic, from the water with plants and animals having higher concentrations of arsenic than freshwater plants and animals. Concentrations of arsenic in organisms are considerably higher than in the water in which they live, but there is little, if any, concentration upward through the food chain [5, 8].

When man is exposed to arsenic, the arsenic is generally distributed to all tissues, but the largest amount can be found in the muscles. Most arsenic is eliminated from the body within 6 days via the kidneys as very little arsenic is excreted in faeces [20]. The association between arsenic ingestion and arsenic found in the urine has led to the monitoring of urine for measuring occupational exposure. In an investigation by Lowry et al. [21], it was found that the arsenic in the tissues was greatest in the protein fraction rather than the acid soluble lipid portion.

Exposure and Toxicity

Smelters are not the principal source for emitting arsenic into the air: the burning of coal is. Burning of coal releases from 0.08–16 μg arsenic per gram of coal [15]. Estimates based on annual consumption of 400 million tons of coal by power plants is approximately 3,000 tons per year [11].

The other principal source of arsenic emission to the atmosphere in the US is cotton ginning dust. Cotton ginning dust and the combustion of cotton gin wastes have been reported as creating significant concentration of arsenic in the air downward from these operations [10]. It has also been reported that a seasonal variation of arsenic concentration in the atmosphere is observed which coincides with the cotton farming activities of harvesting and ginning [3].

The use of arsenic as a medicine, although practiced for hundreds of years reached a peak in the mid to late 1800's. Sowler's solution, containing arsenic trioxide at 10 mg/ml was prescribed for symptomatic relief of many conditions, ranging from acute infections to epilepsy, asthma, and chronic recurring skin eruptions, such as eczema and psoriasis [10]. Many patients received arsenic for months and years and it was in such patients that the consequences of long-term administration of arsenic were first recognized.

Arsenic and all of its compounds are poisonous. Subacute arsenic poisoning usually occurs when a victim is exposed to amounts of arsenic sufficient to cause symptoms but inadequate to make the victim collapse immediately. The victim may go weeks with gradually increasing signs and symptoms related to several organ systems and giving the appearance of a progressive chronic disease state. If death occurs it appears to be due to a natural disease. This appearance has contributed to the popularity of arsenic in homicides.

Acute arsenic poisoning usually occurs through ingestion of contaminated food and drink. The signs and symptoms are variable and depend on the form and amount of arsenic, the age of the patient, and other factors. The major characteristics of acute arsenic poisoning are profound gastrointestinal damage and cardiac abnormalities.

Arsenic enters the human body by inhalation, ingestion, and absorption through the skin. Arsenic content of hair serves as an indicator of exposure to the metal. Upon inhalation of arsenic a mild bronchitis and nasal irritation follows. More concentrated exposures, particularly among those working with arsenic, can lead to perforation of the nasal septum [15]. Skin irritation is also a problem to smelter workers who are exposed to dusts with a high arsenic content.

There is strong epidemiological evidence that inorganic arsenic is a skin and lung carcinogen in man. Lung cancer is associated with inhalation exposure to arsenic in copper smelters, workers in pesticide manufacturing plants, vineyard workers and gold miners [10, 15].

References

1. Anastasia, F.B., Kender, W.J.: J. Environ. Quality 2, 335 (1975)
2. Andreae, M.O.: Analyt. Chem. 49, 820 (1977)
3. Attrep, M., Jr., Anirudham, M.: Atmospheric inorganic and organic arsenic, in: Trace Substances, in Environmental Health-XL [D. D. Hemphill (ed.)] Columbia, Missouri, 1977
4. Calvert, C.C.: Arsenicals in animal feeds and wastes, in: Arsenical Pesticides. ACS Symp. Ser. 7 [E. A. Woolson (ed.)] Washington, D.C.: Amer. Chem. Soc., 1975
5. Conway, H.L.: J. Fish. Res. Bd., Canada. 35 (3), 286 (1978)
6. Ferguson, J.F., Gavis, J.: Water Res. 6, 1259 (1972)
7. Gallorini, M., et al.: Analyt. Chem. 50 (11), 147 (1978)
8. Giddings, J., Eddleman, G.: Water, Air and Soil Pollution 9, 207 (1978)
9. Grosch, M., et al.: Atmos. Environ. 12, 1235 (1978)
10. Nat. Acad. Sci.: Arsenic, Nat. Res. Council, Washington, D.C., 1977
11. Nelson, K.W.: Env. Health Persp. 19, 31 (1977)
12. Rankama, K., Sahama, T.G.: Geochemistry, Univers. Chicago Press, Chicago, Illinois, 1950
13. Richardson, E.W., et al.: J. Env. Quality 7 (2), 189 (1978)
14. Sandberg, G.R., Allen, I.K.: A proposed arsenic cycle in an agronomic exosystem, in: Arsenical Pesticides, ACA Symp. Ser. 7, [E. A. Woolson (ed.)] Washington, D.C., Amer. Chem. Soc., 1975
15. Waldbott, G.L.: Health effects of environmental pollutants, C. V. Mosby, Co., St. Louis, Missouri, 1973
16. Wong, P.T.S., et al.: Methylation of arsenic in the aquatic environment in: Trace Substances, in Environmental Health-XI [D. D. Hemphill (ed.)] Columbia, Missouri, 1977
17. Braman, R.S.: Arsenic in the Environment, in: Arsenical Pesticides, ACS Symp. Ser. [E. A. Woolson (ed.)] Washington, D.C., Amer. Chem. Soc., 1975
18. CRC Press: Handbook of Chemistry and Physics, 59th Ed., Cleveland, Ohio, 1979
19. Porter, E.K., Peterson, R.J.: Biogeochemistry of Arsenic on polluted sites in S.W. England, in: Trace Substances in Environmental Health-XI [D. D. Hemphill (ed.)] Columbia, Missouri, 1977
20. Hunter, F.T., et al.: J. Pharmacol. Exp. Ther. 76, 207 (1942)
21. Lowry, O.H., et al.: J. Pharmacol. Exp. Ther. 76, 221 (1942)

Beryllium

In 1797, while carrying out the chemical analysis of emerald to prove an analgous composition, the French chemist Vauquelin discovered the oxide of a new element, which is called the beryl earth (la terre du beryl), after the name of the analyzed mineral. It was the suggestion of the Editors of Annales de Chimie, who were impressed by the singular taste of the salts, that Vauquelin refer to this compounds as glucine. In Germany the name "Berylerde", a translation of "la terre du beryl" was preferred; however, it was not until 1828 that the name beryllium was widely used and accepted in the literature. The name beryllium is now used except in French chemical literature where glucinium (G) is still used [10].

Production, Use and Shipment

Before beryllium alloys were developed, the principal use of berryllium was as oxide in the manufacture of refractories, spark plugs, high quality electrical porcelains, and as beryllium nitrate in the fabrication of Welsbach gas mantles. It was not until the early 1930's, however, that metallurgical improvements created a large enough demand to justify a "beryllium industry." Berylliumcopper, with small additions has been, by far, the important alloy.

Commercially, beryllium had its beginning in Waterbury, Connecticut, where in 1932, the American Glass Co. rolled beryllium-copper alloys into sheets. Beryllium also received some publicity in 1932 with the discovery of the neutron by Chadwick. This discovery was an outgrowth of the observation that beryllium metal, when bombarded by alpha particles, emits a highly penetrating radiation. In 1936, disks of compressed beryllium powder were used as windows for X-ray tubes because of the transparency of beryllium to soft X-rays. A short-lived use of beryllium oxide was as a phosphor in the fluorescent lamp industry. The use was discontinued in 1949 because of potential health hazards. Beryllium-copper alloys were used widely in World War II for various aircraft engine parts as well as for piston rings and bushings for the Fiat motors in Italy [29].

The world production of beryllium in 1979 was 3000 metric tons, and the US used about 10% of the total [26]. Industrial production before 1968 was from the beryllium aluminosilicate, beryl, which contains about 4.8% beryllium. Since 1968, a hydrous beryllium silicate, bertrandite, has also become an important source.

Beryl ores are processes in one of two ways. They may be finely ground, sintered with sodium fluorosilicate, and the resulting beryllium fluoride leached with water; or the lump ore can be fused, quenched, annealed, and the beryllium sulfate leached with sulfuric acid. In either case, the resulting solutions are purified, and beryllium hydroxide is precipitated. This may be converted to other beryllium compounds, or reduced to the metal. Beryllium metal is then generally broken into small particles that are pressed into masses with various desired shapes through powder-metallurgical methods. Alloys, beryllium metals, and BeO account for all but a small amount of beryllium used [15].

Beryllium and beryl ores are of outstanding interest for nuclear purposes because of their combination of unusual physical and mechanical properties with low neutron capture cross-section and high neutron scattering, i.e., good moderating properties [7]. When beryllium atoms are bombarded with alpha particles from radium, their nuclei disintegrate and yield a profusion of neutrons. Beryllium is one of the most efficient materials for reducing the speed of neutrons, as well as an excellent neutron reflector. The Manhattan Project, overnight, created a large market for beryllium [24]. In the past, beryllium has been used as a moderator and reflector in nuclear facilities such as the American Materials Testing Reactor and the Nautilus Reactor [7].

The principal uses of beryllium stem from the discovery in the 1920's that the addition of only 2% of beryllium to copper forms an alloy 6 times stronger than copper. Beryllium-copper alloys withstand high temperatures, have extraordinary hardness, show resistance to corrosion, do not spark, and are non-magnetic. This alloy is found in many critical moving parts of aircraft engines and in key components of precision instruments, mechanical computers, electrical relays, switches, and camera shutters. Springs made of these alloys retain their springiness almost indefinitely. Beryllium-copper hammers, wrenches, and other tools are employed in petroleum refineries where sparks from steel against stell may cause explosions [24].

Current major uses are aircraft brakes and inertial guidance systems of aircraft and missiles [11]. One example of the types of gyroscopes now possible with beryllium is the electrostatically suspended gyroscope whose rotor spints at 60,000 rpm. After being brought up to speed, if left suspended, the rotor would continue spinning for two years due to absence of frictional forces [4]. Beryllium is also being used in the US Space Shuttle Program for structural components such as the windshield frame, the umbilical door support beams, and brake discs. Other uses include structures on several types of communication satellites produced in the US and Europe. "Lockalloy", a beryllium-aluminum alloy, is currently being used in the ventral fin of a YF12 aricraft and is being considered for use on the X24, an Mach 5 to Mach 9 research vehicle. Ultrathin beryllium foil is finding use in X-ray lithography for reproduction of microminiature integrated circuits, in X-ray image intensifiers, and as windows in radiation detection devices. Fine bore beryllium tubing, now used in quality phonograph style, may eventually find use in high-speed recording devices and dial-type meters [11]. A summary of current uses of beryllium is shown in Table 1.

Chemistry

In the periodic system of the elements, beryllium is the first element of the second group and the third element of the first (helium) period. These circumstances determine the special structural features of the beryllium atom. The atomic weight on the physical scale is 9.05043 and on the chemical scale it is accepted as 9.013 ± 0.0004.

The small radius of the beryllium ion Be^{2+}, the smallest among the metals, strengthens the bond between its valence electrons and the nucleus and is one of the reasons for the high stability of the lattices in beryllium minerals. For this main reason, beryllium has a large electronegativity, which approaches that of aluminum. The chemical properties of beryllium, like those of aluminum, allow it to occupy an intermediate position between the typical cations and the complex-forming elenets. Covalent bonds play an important role particularly in its oxygen compounds. It's atmospheric properties are manifested especially strongly in it's oxides and halides of the alkali metals, in which beryllium plays the role of a complexforming element to form various beryllates [3].

There are five known isotopes of beryllium with mass numbers of 6, 7, 8, 9, 10, of which ^9Be is stable. The nucleus of this naturally occurring stable isotope consists of nine particles (four protons and five neutrons). The odd neutron is not bound in the nucleus as strongly as the other eight particles, so that beryllium may be used as a neutron source. This neutron emission of beryllium due to irradiation may be applied in quantitative estimates of the beryllium content in minerals and ores [30].

Table 1. Uses of beryllium [28]

Form	Use
Berryllium metal	Nuclear applications
	Gyroscopes
	Accelerometers
	Inertial guidance systems
	Rocket propellants
	Aircraft brakes
	Heat shields for space capsules
	Portable x-ray tubes
	Optical applications
	Turbine rotor blades
	Mirrors
	Missile systems
	Nuclear weapons
Beryllium-copper alloys	Springs
	Bellows
	Diaphrams
	Electrical contacts
	Aircraft engine parts
	Welding electrodes
	Non-sparking tools
	Bearings
	Precision casting
	High-strength, current
	Carrying springs
	Fuse clips
	Gears
Beryllium oxide	Spark plugs
	High voltage electrical components
	Rocket-combustion-chamber liners
	Lasar tests
	Electrical furnace liners
	Microwave windows
	Ceramic applications

It is characteristic that beryl deposits occur unpredictably and in isolated pockets. These pockets may be quite large, as in the Las Tapias mine of Argentina which provided 300 tons per year of beryl. Beryl production figures from individual countries are therefore liable to fluctuation as new pockets are discovered and older ones depleted [7].

In the mid 1960's the major world sources of ores that contain beryllium have been India, Brazil, Uganda, Mozambique, Argentina, and the USSR.

Very little beryllium is released to groundwater during weathering as a result of the small amount that escapes capture by growing clay minerals [14]. Under some circumstances, however, beryllium will remain in solution long enough to migrate at least short distances.

Analytical Methods

The gravimetric determination of beryllium may involve the precipitation of beryllium-ammonium phosphate or beryllium hydroxide. In many methods, organic reagents are employed and in others, organic salts of beryllium. Volumetric methods for the determination of beryllium are based on the formation of stable stoichiometric complexes of beryllium. The use of complexes of beryllium eliminates the labor involved in the separation of interfering elements when beryllium is determined in multicomponent media [17].

Photometric methods are widely employed in the determination of microamounts of beryllium. These methods are used primarily when simplicity and speed are preferred. Many colored reagents are employed, as well as reagents such as acetylacetone, salicyclic and sulfosalicyclic acids, whose compounds with beryllium display an intense absorption in the UV range. Colorimetric determination of beryllium in coal has been described by Abernethy and Hattman [1]. This spectrophotometric method is sufficiently sensitive for the determination of microgram quantities and may be desirable for small laboratories without access to spectrographic equipment.

Beryllium may also be determined by polarographic techniques. A sharp polarographic wave of beryllium is obtained in an acid medium (pH 2-4) if LiCl, Li_2SO_4 or tetraethylammonium iodide $(C_2H_5)_4NI$ is used as a background electrolyte, where the diffusion current is directly proportional to the concentration of beryllium.

Also important in the determination of beryllium is the radioisotope ^7Be. It may be obtained carrier-free in the reaction ^7Li (d,2n)^7Be by bombarding a lithium target with deuterone, with subsequent purification by extraction with acetylacetone or theonyltrifloroacetone. ^7Be is used for analytical control in the extraction of small amounts of beryllium, especially from biological materials and in the analysis of meteorites [17].

Spectroscopy is one of the most sensitive techniques for the determination of beryllium. Quantitative and semi-quantitative spectroscopic methods are widely employed in the analysis of industrial and natural including biological materials, and also in the determination of impurities in beryllium and its compounds. Robbins et al. [21] have described the direct determination of beryllium in petroleum and petroleum products at the µg/l level by heated vaporization atomic absorption. This method is applicable to the determination of 1 to 50 ng Be/g with a precision of 10% at the 30–40 ng/g level.

Activation methods for the determination of beryllium are specifically based on two nuclear reactions:

$$^9Be \ (Y,n)^8Be \ \text{ and } \ ^9Be \ (a,n)^{12}C.$$

Compared to other elements, beryllium has a very low neutron binding energy in the nucleus, so that it readily emits neutrons when exposed to x-rays and particles. The threshold of the photo-neutron reaction of beryllium (γ,n) is very low, and the amount of neutrons produced when beryllium samples are irradiated with x-rays is proportional to the content of beryllium in the sample.

Fluorometric methods for the determination of beryllium are based on measurement of the intensity of the fluorescence being given by its compounds with certain organic compounds. The sensitivity of individual fluorometric methods is comparable to that of spectroscopic methods and is much higher than that of spectrophotometric methods. Fluorometric methods also are commonly more selective than spectrophotometric methods [17].

Transport Behavior

During the processes of weathering and formation of sediments, beryllium closely follows the course of aluminum, being enriched, along with aluminum, in clays, bauxites, recent deep sea deposits, and other hydrolyzate sediments. This is due to the similarity of the basic properties and solubilities of these metals. Though radii and charges of the Be^{2+} and Al^{3+} ions differ, their ionic potentials are rather similar ($B^{2+} = 5.9$; $Al^{3+} = 5.3$), and therefore, the two metals follow each other into the hydrolyzates during the hydrolytic decomposition of their salts [19]. Using emission spectroscopy, Durum and Hafty [8] measured beryllium in 59 samples of surface water from 15 rivers in the United States and Canada. The highest concentration observed was less than 0.22 µg/liter.

Accumulation and Metabolism

The ash of many coals have long been known to contain beryllium in concentrations as much as 100 ppm. Since beryllium is mainly in the organic matter of the coal, it is indicative that it was a constituent of the precursor plants.

Modern plants have also been found to contain beryllium. Hickory is the best beryllium accumulator found so far, containing as much as 1 ppm dry weight. Most studies have shown that leaves contain more beryllium than either twigs or fruits, although some desert species contain more beryllium in the twigs. Romney and Childress [22] have shown that beryllium inhibits the growth of several different crop plants at levels above 2 ppm in nutrient solution and 4% of the cation exchange capacity in soil.

It has been shown that the amount of beryllium retained in the body of experimental animals that ingested Be (about 0.006%), was very small compared with control animals, indicating that very little is absorbed by the gastro-intestinal tract. This was attributable to the fact that soluble salts are precipitated by reaction with proteins in the alimentary tract while less soluble salts are not appreciably dissolved in serum or gartric juices [6].

Excretion of beryllium in animals is mainly by the feces – 94% as compared with 1.6% in urine. With intratracheal administration, excretion was always high in the first 24 h and then dropped to low levels which became undetectable after 75 days. Elimination by the feces was persistent for 40 days. The urinary excretion of beryllium in human beings exposed has some outstanding features:

1) There appears to be a quantitative relation to the degree of exposure in individuals not suffering from beryllium poisoning.

2) The rate of excretion is a function of the solubility of the particular compound inhaled.

3) Prolonged excretion can follow even slight exposure: beryllium has been detected in the urine up to 10 years after cessation of exposure.

4) In general the urinary level of beryllium in persons suffering from granulomatosis does not differ from that in persons similarly exposed by unaffected and may not be present in detectable amounts.

Several groups of investigators have reported that beryllium has an inhibitory effect on alkaline phosphatase enzymes. This inhibition brings about widespread cellular changes which, it is believed, may be responsible for the toxic symptoms and signs of chronic beryllium poisoning. It has been suggested that this inhibition may be associated with the displacement of beryllium of magnesium in magnesium-activated enzymes [6].

It appears that storage of absorbed beryllium takes place in the bones rather than in the soft tissues, though there is some transient retention in the liver, kidneys, and lungs. The sites of deposition appear to depend chiefly of the site of exposure and the nature of the beryllium compound, whether soluble or insoluble. It has been found that intramuscular infection of beryllium results in deposition of unexcreted portions of the bone.

Exposure and Toxicity

The following sources, when engaged in operations involving beryllium, are thought to be the most significant sources of beryllium emissions: extraction plants, ceramic manufacturing plants, foundries, machining facilities, propellant manufacturing plants, incinerators, rocket motor test sites, open furning sites for waste disposal. In addition, the emission of beryllium to the atmosphere can occur during the mining of beryllium ores; the improper transportation of beryllium compounds, or wastes contaminated with either; and the burning of oil or coal containing trace elements of beryllium [27, 28]. Another, less localized source of emission, has been shown in mantle-type camping lanterns. The analysis of eight new, unused mantles revealed beryllium levels of about 650 μg. After one hour's use in a lantern, the mantle residues were found to contain about 200 μg. Most of the missing 400 μg of beryllium had volatilized and become airbone in the first 15 min of mantle use. It is likely that many people are unnecessarily exposed to the dangers of beryllium inhalation during the first minutes of use of a new mantle [12]. Beryllium is also used in some dental alloys. When the alloy is ground and polished in a dental laboratory which lacks adequate ventilation, workers may be exposed to beryllium levels two to three times greater than the US Standard ($0.2\ \mu g/m^3$) [5].

It is now known that beryllium can have both acute and chronic effects. On the acute side, it can produce skin ulcers, ocular ulcers, severe pneumonitis, rhinitis, pharyngitis and tracheobronchitis. All of these are reversible if exposure is ended in time. The chronic effects of beryllium, frequently separated by years from the time of exposure, usually include pneumonitis with a cough, chest pain, and general weakness. Pulmonary dysfunction and systematic manifestations including right heart enlargement and congestive heart faulure, enlargement of the liver and spleen, cyanosis (purplish discoloration of the skin and mucous membrane as a result of deficient oxygenation of blood), digital thickening and kidney stones may also appear. Pathologically, chronic effects include interstitial cellular infiltration, granulomatous lesions, and calcific inclusions [2].

Beryllium poisoning, or berylliosis, refers to the pathological condition characterized by pulmonary insufficiency and major changes in the lung. It results from the inhalation of finely divided particles of beryllium metal or beryllium compounds. The first clinical symptoms may be mild, with indisposition, slight by persistent weight loss, weakness, and occasional cough [2].

Beryllium poisoning may be termed acute or chronic, depending upon the duration and intensity of exposure. Acute berylliosis is represented by irritation of the respiratory tract with pneumonitis, and alteration detectable by X-rays. It is commonly caused by shortterm inhalation of heavy concentrations of soluble beryllium compounds. In one reported instance, the beryllium fluoride concentration in the air had risen to 450–600 $\mu g/m^3$ beryllium for 20 min. Three individuals inhaled about 59 mg and developed the acute disease within 72 h. Onset of the acute disease is rapid, lasts for weeks, and, if not fatal, is followed by apparent complete recovery [7].

The symptoms of chronic berylliosis are less clearly defined. Loss of weight, coughing, weakness and breathlessness from little exertion are characteristic. Bone and lung sarcomas have been observed in animals, but not in humans. The observed symptoms may not appear for as many as 15 years. The illness is of long duration; it may be completely incapacitating, and mortality is high. Attempts to estimate the beryllium concentrations responsible for this disease are complicated by the so-called "neighborhood cases." In 42 cases of this type reported prior to 1960, 50% were fatal and were associated with people who lived within three-fourths of a mile of a beryllium plant, and exposed to concentrations of only 0.01–0.1 $\mu g/m^3$. In one instance, there were more neighborhood cases of poisoning than cases among workers in the plant who were exposed to higher concentrations. It has been suggested that at some distance from the plant, the beryllium-bearing particles are smaller and therefore more toxic than at the plant itself [7]. In fact, one case pertained to a woman who lived two miles from the beryllium plant and received her exposure from the dust from the clothes of her husband, who worked in the plant [24, 25].

New cases of beryllium poisoning are still appearing. Over 80 cases have been reported to the Beryllium Case Registry between 1966 and 1974, of which 36 patients were exposed to beryllium after 1949, a time when industrial controls had been established. Most of these patients were exposed in the aviation and nuclear industries [13].

There is no direct evidence that beryllium is carcinogenic to humans, but the fact that neoplasms of varying types have been produced in the lungs of rats by inhalation, and osteosarcomas in other animals by intravenous injection and inhalation, has given rise to some speculation to the potential effect. One of the most impressive findings of Schepers et al. [23] in their inhalation experiments was the large number of lung neoplases, some of which were malignant, after prolonged inhalation of an aerosol containing beryllium. More recent investigations by Reeves et al. [20] concerning inhalation exposure of rats to beryllium sulfate aerosols have revealed tumors which appeared to be alveolar adenocarcinomas. Dultra and Roth [9] succeeded in producing an osteosarcoma in a rabbit which had been exposed to beryllium oxide dust at a concentration of 6 µg/l for 11 months. The animal died 17½ months after the last day of exposure. About one-eighth of the total substance of the lungs was occupied by spherical masses of metastases, as well as some in the parietal pericardium, spleen, and liver.

At the current time, beryllium-induced carcinogenicity in humans is debatable. However, a study conducted by the US National Institute for Occupational Safety and Health, known as Bayliss III, has called for a reduction in the acceptable limits of beryllium as a result of animal data obtained in 1975. Bayliss III claimed that on epidemiological grounds humans exposed to beryllium incur a risk of lunc cancer [16].

It is apparent that beryllium is an integral component of a large number of manufactured products, particularly with respect to aerospace applications. However, since beryllium has been positively correlated with various pulmonary disorders, stringent emission standards should certainly be implemented. Also, as a result of the association of beryllium with cancer in animals, a continuing effort should be made to study the toxicology of beryllium through epidemiology and chronic low-level feeding studies.

References

1. Abernethy, R.F., Hattman, Elizabeth A.: Colorimetric Determination of Beryllium, in: Coal, US Department of the Interior, Bureau of Mines. Washington, D.C. 8.p., 1970
2. Anonymous: Beryllium, Environment *16*, no. 3. (1974)
3. Beus, A.A.: Beryllium, W.H. Freeman, San Francisco, p. 161, 1962
4. Bleymaier, J.S., Wiese, M.: Systems Considerations for Aerospace Beryllium Applications, Aeronautics and Space Engr., Manufacturing Meet., Los Angeles, CA., Oct. 7–11, 1968, p. 23
5. Brody, J.E.: Beryllium Camping Lantern Mantles and Dental Alloys Called Possible Peril, The New York Times, Apr. 9., p. 25, 1974
6. Browning, E.: Toxicity of Industrial Metals, 2nd edit., Butterworth, London, p. 383, 1969
7. Darwin, G.E., Buddery, J.H.: Metallurgy of the Rares Elements – 7. Beryllium, Butterworths, London, p. 392, 1960
8. Durum, W.H., Haffty, J.: Occurrence of Minor Elements in Water, US Geological Survey Circular 445, Washington, D.C., 1961
9. Dultra, F.R., Roth, J.L.: A.M.A. Arch. Path. *51*, 473 (1951)
10. Everest, D.A.: The Chemistry of Beryllium, Elsevier, New York, p. 151, 1964
11. Farkas, Martin S.: Engr. Mining *178*, 169 (1977)
12. Griggs, K.: Science *181*, 842 (1973)
13. Hasan, F.M., Kazemi, H.: Chest. *65*, (no. 3), 289 (1974)
14. Kopp, J.E., Kroner, R.C.: Trace Metals in Waters of the United States. A five year summary of trace metals in rivers and lakes of the United States. (Oct. 1, 1962—Sept. 30, 1967). US Dept. of the Interior. Federal Water Pollution Control Administration, Div. Poll. Surveillance, Cincinnati, Ohio, 1967
15. Nat. Acad. Sci., Nat. Res. Council, US Committee for Geochemistry. Geochemistry and the Environment, vol. II, The Relation of other selected trace elements to health and disease, Washington, D.C., 1973
16. Nat. Acad. Sci., Nat. Res. Council, Assembly of Life Sciences, Committee on Toxicology, Drinking Water and Health, Washington, D.C., p. 939, 1977
17. Novoselova, A.V., Batsanova, L.R.: Analytical Chemistry of Beryllium, Ann Arbor-Humphrey Science Publishers, Ann Arbor, p. 225, 1969

18. Pattee, E.C., Van Noy, R.M., Weldin, R.D.: Beryllium Resources of Idaho, Washington, Montana, and Oregon, US Dept. Interior, Bureau of Mines, Washington, D.C., p. 169, 1968
19. Rankama, K., Sahama, T.G.: Geochemistry, Chicago Press, Chicago, Illinois, p. 912, 1950
20. Reeves, A.L., Deitch, D., Vorwald, A.J.: Cancer Res. *27*, 439 (1967)
21. Robbins, W.K., Runnels, J.H., Merryfield, Ruth: Analyt. Chem. *47*, 2095 (1975)
22. Romney, E.M., Childress, J.D.: Soil Sci. *100* (1965)
23. Schepers, G.W.H., Durkan, T.M., Delephant, A.B., Creedon, F.T.: A.M.A. Arch. Ind. Hlth. *15*, 32 (1957)
24. Schubert, J.: Beryllium and Berrylliosis, *199*, (no. 2), 27 (1958)
25. Schubert, J., Rosenthal, M.W.: Arch. Indust. Hlth. *19*, 16 (1959)
26. US Department of the Interior. Minerals Yearbook, 128.37: 978–979/vl. 1980
27. US Environm. Protect. Agency. Bat, Inventory of Sources and Emissions, Beryllium-1968, NTIS PB-220975, p. 42, 1971
28. US Environm. Protect. Agency, Office of Air and Water Programs, Control Techniques for Beryllium Air Pollutants, Research Triangle Park, North Carolina, PB-AP 116, p. 28, 1973
29. White, D.W., Burke, J.E.: The Metal Beryllium, Amer. Soc. Metals, Cleveland, Ohio, p. 703, 1955
30. Vlasov, K.A.: Geochemistry and Mineralogy of Rare Elements and Genetic Types of Deposits, vol. 1, Israel Program for Sci. Tranlations, Jerusalem, p. 688, 1966

Selenium

Selenium was discovered in 1817 by Berzelius, a Swedish chemist. He was searching for tellurium in flue dust from lead chambers at a sulfuric acid plant. Berzelius noticed that the reddish color sludge in these chambers became malodorous upon heating and eventually isolated the cause, selenium. He chose the name from the Greek word for Moon, selene, because of the similarity to tellurium, whose name is derived from the Latin tellus, meaning Earth [44].

For a number of years selenium remained a chemical curiosity. In 1873 Smith discovered the unique electrical properties of selenium and brought it to public attention [7]. In the 1930's selenium was identified as the cause of "blind staggers" and "alkali disease" in livestock and it became of interest because of its toxic effects [47]. The role of selenium as an essential element was proposed in 1957 when Schwartz and Foltz [42] determined that selenium was necessary to prevent liver necrosis in rats. Subsequent studies showed selenium was an essential trace element for several animals [32].

The importance and understanding of selenium's role in nutrition, and the development of industrial uses have made selenium an important metalloid and it is now a significant element industrially, biologically and also environmentally.

Production, Use and Shipment

The western world figures for selenium production from 1964 to 1973 average 1042 tons [33]. The 1976 figure was only slightly higher at 1087 tons. The US produced 184 tons and imported the remaining portion of the 453 tons used from Canada (42%), Japan (27%) and others.

Selenium is used widely in industrial products and processes with the glass industry using 20% of the total US consumption. Selenium is used by glass makers to counteract coloration due to iron oxides. Selenium itself colors glass red if added in sufficient amounts and the environmental glass and other tinted glass used in construction are made using selenium [33].

Pigments made with selenium are used in plastics, paints, and ceramics because they are resistant to heat, light, weather and chemicals. The rubber industry uses selenium additives to increase heat and wear resistance, and to improve elasticity [53]. The strong antioxitant qualities fo selenium make it useful for inclusion in inks, mineral and vegetable oils, and lubricants [33]. Selenium is valuable as a general catalyst for the production of organic and petrochemicals (53) and it is used as a catalyst in the production of pharmaceuticals such as niacin and cortisone [33].

Selenium was used in pesticides, being especially effective as a miticide. Its use has been restricted to green house ornamentals because of the threat of carcinogenicity [45].

Electronic uses of selenium are related to its semiconductor and photoelectric characteristics. Selenium is used in the manufacture of semiconductors, thermoelements, photoelectric and photocells, and xerographic materials. In metallurgy selenium improves the stability and machineability of various allys, is used to color copper alloys, and is added to stainless steel to enhance the corrosion resistance [44].

Since the establishment of selenium as an essential element, selenium has been used as a feed additive for livestock. Small amounts (0.1–2 ppm) have been added to feeds since 1974 when the FDA lifted the ban on selenium [33]. Natural source selenium tablets are now available for human consumption.

Medical uses of selenium are restricted to topical applications for treatment of dandruff. These compounds are in the form of selenium mono- and disulfide with Selson Blue (1% mixture) and Selsun Red (2.5 mixture) as the major products. Selenium sulfides are also used to treat *Tinea versicolor*, a major fungal infection [33].

Limited use of radioactive selenium (^{75}Se-selenomethionine) as a diagnostic scanning and labeling material is being investigated because these compounds concentrate in the liver, pancreas, and other important organs which are difficult to study with x-rays [8, 33, 40].

Selenium shows a strong affinity for sulfur and copper, and most commercial operations are associated with sulfur or copper deposits [44]. Selenium occurs as a minor element in the copper areas in the US and Canada, and it is also associated with the copper-nickel-cobalt ores of Eurasia. While most of the selenium is produced in North America, significant deposits are found in Asia, Africa and South America. The largest deposit is located at Pacajake in Bolivia [44].

The distribution of selenium in soils is primarily in weathered products of sedimentary formations containing selenium. Padalfers contain varying amounts of elemental and selenide forms, while pedocals of arid areas are characterized by high levels of oxidized forms. In the US the highest soil levels are found in areas of the west and midwest, particularly in soils derived from Cretaceous beds [27].

Chemistry

The chemical and physical properties of selenium are intermediate between metal and non-metal. This is explained by its position in group VIA of the periodic chart,

Table 1. Main properties of selenium [53]

Property	Selenium
Atomic number	34
Atomic mass	78.96
Density, g/cm^3	4.79 [a]
Melting point, °C	217
Boiling point, °C	685.4
Atomic radius, mu	0.117
Hardness, relative units	2 [a]
Electronegativity, relative units (Li = 1)	2.4
Ionic radius, mu	
E^{2-}	0.198
E^{4+}	0.069
E^{6+}	0.035
Latent heat of fusion, joule/g (cal/g)	6.91 (16.5)
Heat of vaporization, joule/g	272.98 (65.2)
Specific heat C_{solid}, kjoule/kg deg (cal/kg deg)	0.318 (0.076)
Specific heat C_{liq}, kjoule/kg deg (cal/kg deg)	0.494 (0.018)
Specific heat at constant pressure, joule/g deg	0.360 (0.086) [a]
Thermal conductivity, W/m deg	0.293–0.766

[a] Hexagonal modification

between sulfur and tellurium. Elemental selenium is similar to tellurium, while in combination with other elements it has characteristics similar to sulfur. Some of the basic physical properties of metallic selenium are shown in Table 1.

Pure selenium, like sulfur, is allotropic, and exists in three forms; a gray hexagonal form, a red monoclinic form, and a vitreous amorphous form. All forms can exist at room temperature and one atmosphere, but only the hexagonal form is thermodynamically stable [53].

An interesting physical feature of the hexagonal form of selenium is its photoconductivity. Conductivity is low in the dark, but increases 1000 fold when subjected to light energy [53]. This is caused by the emission of electrons prompted by light energy [44]. The elemental form also becomes a better conductor when heated.

There are six natural isotopes of selenium which exist under normal conditions.

Isotope	^{74}Se	^{76}Se	^{77}Se	^{78}Se	^{80}Se	^{82}Se
Abundance	0.87	9.02	7.58	23.52	49.82	9.19

In addition, there are ten, short-lived man-made isotopes of which ^{75}Se, ^{77m}Se, and ^{81m}Se are the most used in neutron activation, radiology, etc. The isotope used in medical and biological tracer studies, is ^{75}Se usually in the form of ^{75-}Se selenomethionine [33].

Selenium has a valence of 2- when combined with hydrogen or metals. In oxycompounds it has 4+ and 6+ valence, similar to sulfur. Like sulfur, selenium forms dioxides, trioxides and other oxygen compounds, especially if heated. The most common oxidation reactions are conversion of elemental selenium to selenium dioxide, which dissolves in water to form selenious acid (H_2SeO_2). The presence of oxidizing agents may transform selenious acid to selenic acid. Both of these acids have properties similar to the corresponding acids of sulfur.

Molten selenium mixed whith metals will produce either metal selenides for alloys, most of which are highly insoluble, particularly those of the heavy metals. Mercury selenide, for example, has a K_{sp} of 10^{-59} [33]. Most of the organic and inorganic complexation chemistry involve the 2-oxidation state, and many of the compounds are sulfur analogs, where selenium has displaced sulfur. An exhaustive treatment of the selenium complexation chemistry is found in Nazarenka and Ermakov [53].

Analytical Methods

A variety of qualitative and quantitative methods of analysis have been developed for selenium. Qualitative determinations are fairly straightforward, but quantitative techniques are hampered by losses due to volatilization. This is especially critical because the levels being detected must often be submicrogram, and even small losses yield great error. Quantitative tests involve reduction to elemental selenium with an accompanying precipitate formation or color change. A summary of the common qualitative tests is given in Table 2. For quantitative analysis, gravimetric methods, redox titrations and several instrumental techniques are used. A complete treatment of the titration and gravimetric methods for selenium is found in Nazarenko and Ermakov [53] and will not be covered here.

Table 2. Qualitative determinations for selenium [53]

Reagent	Valence state of element	Result of reaction	Detection limit, $\mu g/ml$	Interfering elements
Thiocyanic acid	Se(IV)	Red brown precipitate	2	As, Sb, Sn, Fe(II), MoO_4^{2-}
Pyrrole	Se(IV)	Pyrrole blue	0.5	Oxidizing agents; Se(VI), Te(IV), Te(VI)
Assymmetric diphenyl-hydrazine	Se(IV)	Red Color	2	Oxidizing agents; Te does not interfere
Methylene blue and sodium sulfide	Se°	Decolorization of methylene blue	3	Oxidizing agents
Ammonia molybdate	Se(IV)	Molybdenum-selnium blue	3	PO_4^{3-}, SO_4^{2-}
3,3′ Diaminobenzidine	Se(IV)	Yellow color or red fluorescence	2	Oxidizing agents; Fe(III), Cu(II)
2,3 Diamononaphthalene	Se(IV)	Ditto	2	Oxidizing agents
Thiourea	Se(IV)	Pink color or red precipitate	5	Te, NO_2^-, Cu, Hg, Au, Pt, Pd, Bi
Hydroxylamine hydrochloride	Se(IV)	Ditto	5	Many elements; Te does not interfere
Iodide	Se(IV)	Red brown precipitate	40	As(III), Ge(IV), Mo(VI), Te does not interfere

Photometric and fluormetric analysis is based on the measurement of the following products of selenium reactions:

1. Hydrosols
2. orthodiamines
3. colored azo compounds made from diazonium salts produced by oxidation with selenium.
4. complexes of selenium with sulfur containing compounds
5. complexes of selenium(II) with phenyl-substituted semicarbazides and thiocarbazides.

Hundreds of papers cover the modifications and improvements of the various chemical treatments [3, 9, 14, 16, 50, 51]. A complete review is found in Nazarenko and Ermakov [53].

The most important methods involve reaction of selenium with aromatic orthodiamines to form colored piazoselenols. These reactions are the basis for most of the current standard methods of selenium determination [46]. In addition to spectrophotometric analysis, the major instrumental methods include neutron activation, atomic absorption spectroscopy, x-ray fluorescence and gas chromatography.

Neutron activation is the standard against which other techniques are measured because of its accuracy and sensitivity. ^{75}Se is the most commonly used isotope [41], but a new technique developed by McKnown and Morris [30] uses ^{m77}Se and requires only 20 min per sample.

Flame atomic absorption spectroscopy has the disadvantage of low lamp intensities and poor stability, but modifications of sample treatment and use of special gas combinations improve selenium detection [24, 39]. Flameless techniques have shown more promise [26, 29] and the detection limit of 72 picograms obtained by Baird et al. [5] shows that the necessary sensitivity is attainable.

X-ray fluorescence is applicable to selenium analysis for biological and other samples, and show good correlation with neutron activation [20, 52]. The volatile nature of seleno-compounds makes them good subjects for gas chromatography. An advantage of gas chromatography is the available field units which can detect levels as low as 0.1 ppm [18].

Transport Behavior

Although selenium levels are low in natural waters, fluvial action is important in selenium transport. Bectine and Goldberg [54] estimated that 8,000 tons of selenium are deposited annually in the oceans. The average concentration of selenium in seawater is 0.09 ppb, due to loss by precipitation of selenite-metal hydrozies. Analysis of rainwater for selenium content [21, 25] have shown a range of 0.04 to 1.4 ppb, and atmospheric selenium in rainwater is correlated to industrial emissions [21].

Studies of biological selenium conversions show that microorganisms can metabolize selenium. Chau et al. [12] found that several types of aquatic bacteria converted seleno-compounds to volatile methylated forms. In nature, all valence forms of selenium exist. These forms depend on the solubility and oxidation reduction reactions possible in the environment. Figure 1 illustrates the pH and pE relation-

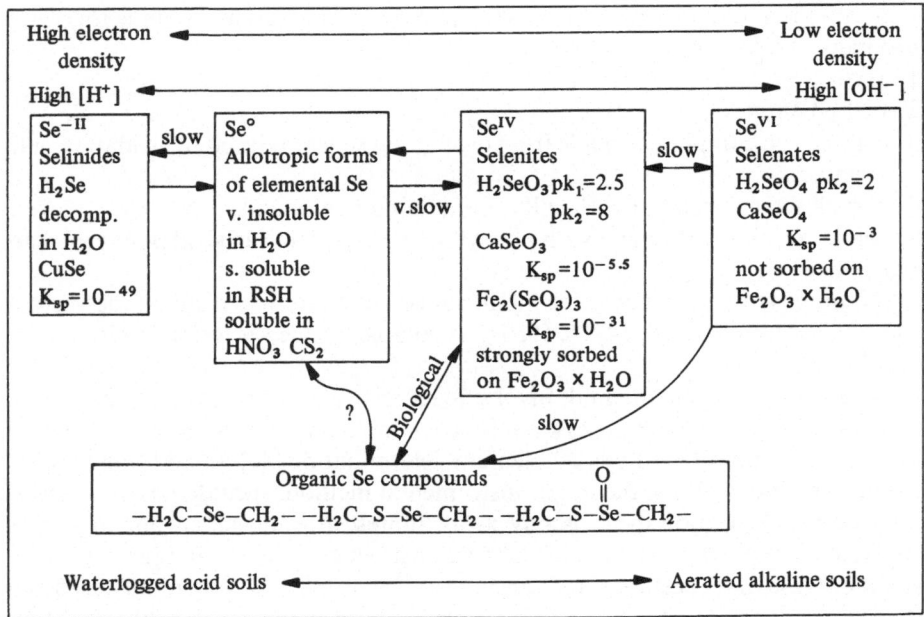

Fig. 1. Generalized chemistry of selenium in soils and weathering sediments [3]

ship to chemical form and show the various chemical forms occurring in soil and sediments for a range of conditions.

Most selenium occurs as insoluble elemental and selenide forms, but in oxidizing environments it is converted to selenites and selenates, both of which are soluble [3]. These soluble oxyanions are washed into waterways or leached back down to reducing zones where they again become insoluble. In arid areas, with alkaline soils, selenate is a predominant and stable form [27]. Elsewhere, both selenate and selenite are easily reduced. Selenite forms have strong affinity for metal hydrozies, and are adsorbed onto insoluble complexes under acidic soil conditions or in water [22]. The high selenium content of some Hawaiian soils of the earth's crust is estimated at 0.09 to 0.2 ppm. It is the 40th most common element, between bismuth and gold [7].

The ionic radii of Se^{2-} and S^{2-} are similar, and selenium often displaces sulfur in minerals [7]. There are approximately 40 minerals in which selenium predominates [53]. An extensive work covering all known selenium minerals has been complied by Sindeeva [44].

Environmentally, selenium has become more important as its use has increased in industry. Selenium water pollution is minimal due to rapid conversion to insoluble selenite metal hydrozies in natural and waste water [3]. Very little data is available on the discharge of selenium in liquid waste. Johnson [23] determined that stack and quench waters in industry contained approximately 0.014 mg/l selenium.

Accumulation and Metabolism

The occurrence of selenium in plants has been known since 1932 and thousands of plant analyses have been carried out on crop, forage and range plants [6]. Selenate is the form considered most available to plants [49], although there may be uptake of other forms by accumulator species.

Some plants concentrate selenium at levels in excess of 1,000 ppm, even on soils with less than 1 pmm [6] while others do not accumulate appreciable amounts even on seleniferous land [49]. Crop plants and non-accumulators tend to incorporate selenium as protein-bound selenomethionine [38, 47] while accumulators take up selenium and convert it to soluble inorganic forms and several free amino acids, notable Se-methylselenocysteine and Se-selenocystathionine [13, 29, 36].

Many plants and algae convert selenium into volatile compounds such as methyl and dimethyl selenides. This may be a mechanism for reduction of selenium to tolerable non-toxic levels [43]. Some accumulator plants have a strong odor [6] caused by these volatile forms, and large amounts of selenium are lost upon drying [33].

The forms of selenium in plants are better known than their biochemical formation. Most of the organic forms are sulfur analogs and the pathways may be similar to sulfur compound formation [47]. Nisson and Benson [37] suggested the following pathway for reduction of selenate to selenite, which is the form which reacts to form amino acids containing selenium.

$$SeO_4 \xrightarrow{\text{sulfate adenyl transferase}} \text{adenosine } 5'\text{-phosphoselenate}$$

$$\text{reductase} \downarrow \text{system}$$

$$SeO_3^{2-}$$

A general scheme for the biochemical pathways of selenium in plants is illustrated in Fig. 2, along with the flow of various selenium compounds between plants, animals and soils.

Accumulator plants occur primarily on highly seleniferous soils and have the capacity to concentrate selenium at levels of 100 to well over 1,000 ppm. Levels of 4,000 ppm have been recorded [6]. Selenium may be essential for some accumulators, but is detrimental to most nonaccumulators at high levels [11]. *Astragalus* (locoweed) is the genus with the most accumulator members [6], but not all species of *Astragalus* have this capacity as shown in Table 3. Other strong selenium accumulators are Prince's Plume *(Stanleya sp)*, and species of *Machaeranthera* and *Haplopappus* (Camphor daisies). Minor accumulators (30 to 100 ppm) include several species of *Aster, Atriplex, Castelleja, Grindelia* and others [6, 33].

These accumulators are characteristic of highly seleniferous soils and can be used as indicators of such areas. For this reason they are used in geologic prospecting for ores of uranium and vanadium which are often associated with selenium [10]. Figure 3 shows the US distribution of selenium in vegetation and in particular the sites where accumulator plants occur.

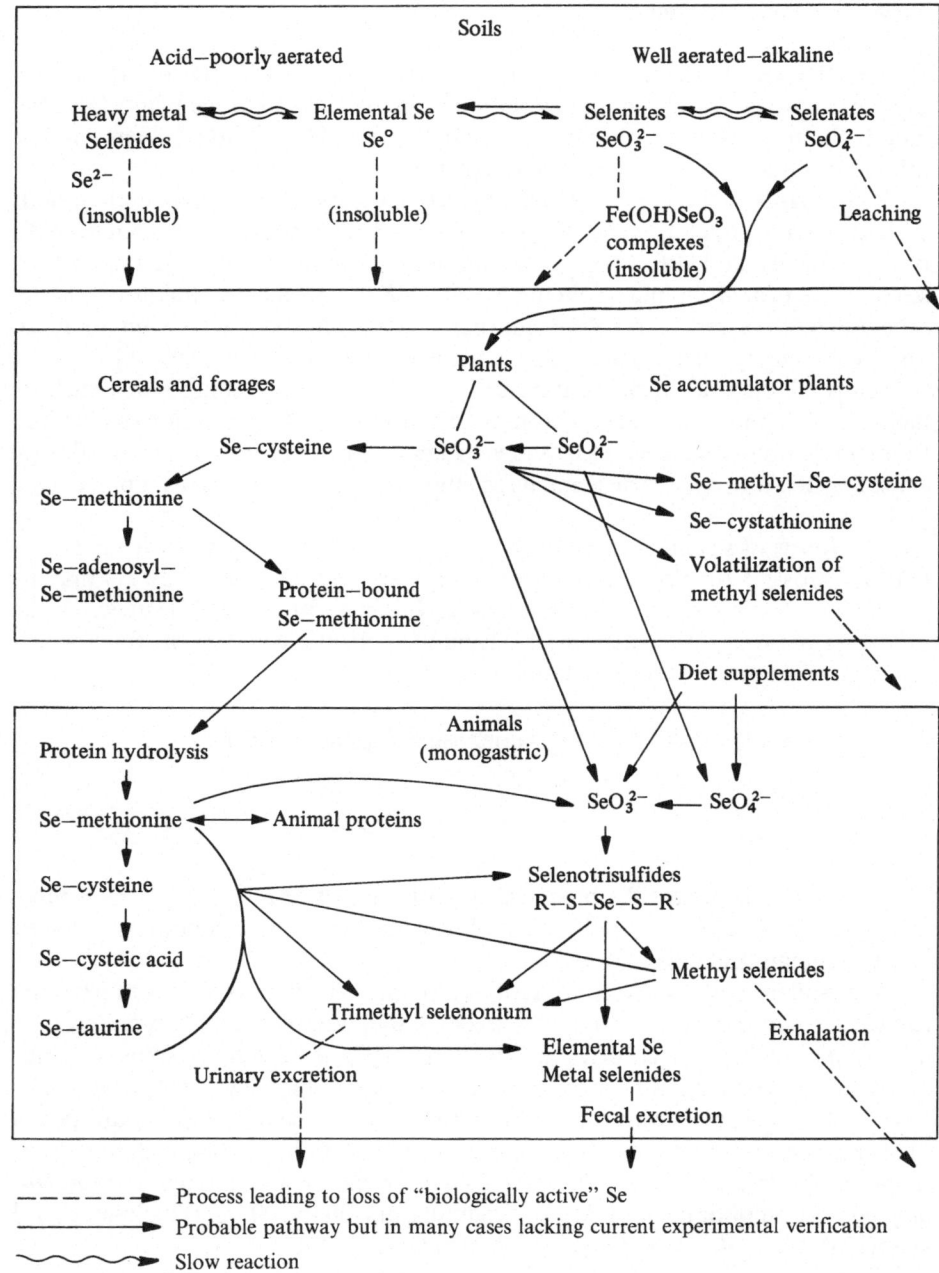

Fig. 2. Chemical and biochemical changes in selenium possibly involved in its movement from soil through plants and animals [33]

Table 3. Variation of accumulation of selenium in the genus *Astragalus.*
(Adapted from [49])

Species	Se content	Soil content	Locale	Ref.
A. racemosus	450	2–14	NE	[5]
A. missouriensis	5			
A. racemosus	5,560	5	NE	[5]
A. missouriensis	25			
A. racemosus	400	0.7	NE	[5]
A. missouriensis	25			
A. bisulcatus	2,620	1.6	N. Dak.	[14]
A. caryocarpus				
A. bisulcatus	2,050	2	Wyoming	[4]
A. missouriensis	5			

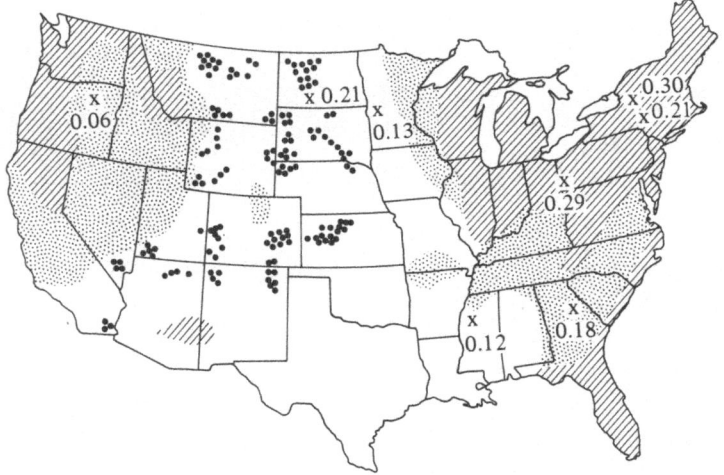

▨ Low–approximately 80% of all forage and grain contain <0.05 ppm of selenium

▦ Variable–approximately 50% contains >0.1 ppm

☐ Adequate–80% of all forages and grain contain >0.1 ppm of selenium

• Local areas where selenium accumulator plant contain >50 ppm

x Selenium concentration in rainwater (µg/l) by sampling location

Fig. 3. Selenium levels in forage, crops, and accumulators and their location in the United States [25]

Exposure and Toxicity

Acute toxicity occurs in both livestock and laboratory animals. The specification of "lethal levels" is difficult because a variety of factors affect toxicity, such as the method of administration, form and concentration of selenium, and the animal

Table 4. Acute toxicity of some selenium compounds [33]

Compound	Experimental animal	Mode of administration	Toxicity
Sodium selenite	Rat	Intraperitoneal injection	MLD 3.25–3.5 mg Se/kg body wt
	Rat	Intravenous injection	MLD 3 mg Se/g body wt
	Rabbit	Intravenous injection	MLD 1.5 mg Se/kg body wt
	Rat	Injection	MLD 3–5.7 mg Se/kg body wt
	Rabbit	Injection	MLD 0.9–1.5 mg Se/kg body wt
	Dog	Injection	MLD 2 mg Se/kg body wt
Sodium selenate	Rat	Intraperitoneal injection	MLD 5.5–5.75 mg Se/kg body wt
	Rat	Intravenous injection	MLD 3 mg Se/kg body wt
	Rabbit	Intravenous injection	MLD 2–2.5 mg Se/kg body wt
Selenium oxychloride	Rabbit	Application to skin	83 mg of compound caused death in 5 h; 4 mg caused death in 24 h
Hydrogen selenide	Rat	In air	All animals exposed to 0.02 mg/liter of air for 60 min died within 25 days
DL-selenocystine	Rat	Intraperitoneal injection	MLD 4 mg Se/kg body wt
DL-selenomethionine	Rat	Intraperitoneal injection	MLD 4.25 mg Se/kg body wt
Diselenodipropionic acid	Rat	Intraperitoneal injection	LD_{50} 25–30 mg Se/kg body wt
Dimethyl selenide	Rat	Intraperitoneal injection	LD_{50} 1,600 mg Se/kg body wt
Trimethylselenonium chloride	Rat	Intraperitoneal injection	LD_{50} 49.4 mg Se/kg body wt

species, age and condition. Table 4 illustrates these variations in the determination of toxic levels in laboratory animals.

Injection of selenium compounds at 200 µg/kg body weight had caused acute toxicity in horses, cattle and swine, and ingestion of forage containing 400 to 800 ppm of selenium is lethal to sheep, calves and pigs [33]. Symptoms of acute selenium poisoning always involve the vascular system. Temperature is elevated, the pulse is rapid and weak, and there is labored breathing. Death is caused by anoxia [33]. The range disease "blind staggers" is a form of acute selenium toxicity [48]. Laboratory animals have shown various degrees of damage to the internal organs, vascular edema and hemorrhage when exposed to selenium in food, water and vapor forms [33].

Acute toxicity in humans is rarely seen because of the low toxicity of elemental selenium used in industry [33]. Toxicity in humans causes the same symptoms as in animals; pulmonary edema, hemorrhaging and vascular disruption [33]. Estimates of acute levels are lacking, although a level of 0.2 grams of sodium selenate is considered acutely toxic [17].

Chronic toxicity in animals and humans is more common. In livestock it is commonly caused by ingestion of vegetation containing 3 to 20 ppm of selenium for prolonged periods [32]. The result is loss of vitality, lameness, elongated and disfigured hooves, degeneration of the internal organs, and loss of hair [48]. Involvement of the hair, nails and hooves suggests that selenium displaces sulfur in keratin, causing structural deformation. Alkali disease is a form of chronic selenium poisoning.

Laboratory animals show similar symptoms, and a dietary level of 5 to 10 ppm has been found to induce chronic toxicity in rats [32]. The vascular system and liver are the most affected by long term exposure to selenium.

Chronic toxicity in humans is usually seen only in industry [17] although cases of chronic poisoning in seleniferous areas have been studied [45] and attributed to consumption of food contaminated with selenium [15]. The daily uptake of 1 mg of selenium/kg body weight may produce chronic toxicity in humans and 5 ppm in foods or 0.5 ppm in water or milk is considered potentially dangerous [17]. Chronic human toxicity is usually caused by inhalation or skin absorption in industry. Symptoms include respiratory inflammation, pulmonary edema, dermatitis, and gastrointestinal congestion and hemorrhage.

There is a metallic taste in the mouth, a noticeable garlic breath odor and very painful inflammation of the nail beds [33]. Long term effects may involve cirrhosis of the liver, kidney damage and spleen atrophy, but little information is available [17].

Selenium has also been implicated as a cause of dental caries [19] and has been shown to be teratogenic [48]. An early study by Nelson [35] showing selenium to be carcinogenic has never been solidly substantiated by subsequent studies [33].

Beneficial aspects of selenium started with its establishment as an essential trace element for animals. Schwartz and Foltz [42] found selenium prevented liver degradation in rats, and a number of subsequent studies showed that it was responsible for the prevention of myopathic diseases in poultry and sheep [31, 34]. Selenium is the third factor, in addition to vitamin E and cystine, that prevents myopathy [32], and it may be a key factor in the absorption of vitamin E. Most selenium is adequately supplied through diets, although some areas such as Oregon and New Zealand have selenium deficient forage and provide supplementary selenium in the feed. Levels below 1 ppm are considered safe, and the required level is estimated to be 0.2 ppm [32].

The use of selenium has remained relatively constant. A pollution problem with the copper mining operations, which gear up or shut down depending on the economy situation [1] is something to be considered for the future. The newest uses, as feed supplement and in the making of "environmental flat-glass" for solar heat barriers should increase. The bulk of the investigations thus far indicate that selenium is not a serious pollutant in most instances, and further research into its importance as an essential trace element my uncover other useful information about selenium.

References

1. Anonymous: Eng. Min. J., March, 1977
2. Anonymous: ibid. Feb, 1979
3. Allaway, W.H.: Control of the environmental levels of selenium, p. 181–206, in: [D. D. Hemphill (ed.)] Trace Substances in Environmental Health – II, Proc. Univers. Missouri's 2nd annual Conf. Trace Substances in Environm. Health, Columbia, Univers. of Missouri, 1968
4. Allaway, W.H., Cary, E.E.: Analyt. Chem. *36*, 1359 (1964)
5. Baird, R.B., Pourian, S., Gabrielian, S.M.: ibid, *44*, 1887 (1972)
6. Beeson, K.C.: Occurrence and significance of selenium in plants, p. 34–40, in Agr. Handbook 200, US Dept. Agr., Washington, D.C., 1961

7. Bhappu, R.B.: Economic recovery of selenium by flotation from sandstone ores of New Mexico, State Bureau of Mines and Mineral Resources, New Mexico Inst. Mining & Technology, 1961
8. Blau, M., Bener, M.A.: Radiology *78*, 947 (1962)
9. Bonhorst, C.W., Mattice, J.J.: Analyt. Chem. *31*, 2106 (1959)
10. Brooks, R.R.: Geobotany and Biogeochemistry in Mineral Exploration, Harper and Row, New York, 1972
11. Broyer, T.C., Johnson, C.M., Huston, R.P.: Plant Soil *36*, 635 (1972)
12. Chau, Y.K., et al.: Science *192*, 1130 (1976)
13. Chow, C.M., Nigan, S.N., McConnell, W.B.: Phytochemistry *10*, 2693 (1971)
14. Cummins, L.M., Martin, J.L., Maag, D.D.: Analyt. Chem. *37*, 430 (1965)
15. Dudley, H.C.: Selenium as a potential industrial hazard, US Public Health Rep. *53*, 281 (1938)
16. Dye, W.B., et al.: Analyt. Chem. *35*, 1687 (1963)
17. Fishbein, L.: Toxicology of selenium and tellurium, p. 191–240, Adv. Modern Toxicology, Hemisphere Publ. Corp., Washington, 1972
18. Gosink, T.A.: Env. Sci. and Tech. *9*, 630 (1975)
19. Hadimarkos, D.B.: Trace elements and dental health, p. 25–30, in: [D. D. Hemphill (ed.)] Trace Substances in Environmental Health-VII. Proc. Univers. Missouri's 7 th Annual Conf. Trace Substances in Environm. Health, Columbia, Univers. Missouri, 1972
20. Handley, R.: Analyt. Chem. *32*, 1719 (1960)
21. Hashimoto, Y., Hwang, J.Y., Yanagisawa, S.: Environ. Sci. Tech. *4*, 157 (1970)
22. Howard, J.H.: Control of geochemical behavior of selenium in natural waters by absorption on hydrous ferric oxides, p. 485–496, in: [D. D. Hemphill (ed.)] Trace Substances in Environmental Health-V., Proc. Univers. Missouri's 7 th Ann. Conf., Trace Substances in Environm. Health, Columbia, Univers. Missouri, 1973
23. Johnson, H.: Environ. Sci. Technol. *4*, 850 (1970)
24. Kahn, H.L., Schallis, J.E.: Atomic Absorption Newslett. *7*, 5 (1968)
25. Kubota,, J., Cary, E.E., Gissel-Neilsen, C.: Selenium in Rainwater of the US and Denmark, p. 123–130, in: [D. D. Hemphill (ed.)] Trace Substances in Environmental Health-IX., Proc. Univers. Missouri's 9 th Ann. Conf. Trace Substances in Environm. Health, Columbia, Univers. Missouri, 1975
26. Kunselman, G.C.: Atomic Absorption Newslett. *15* (2), 29 (1976)
27. Lakin, J.W.: Geochemistry of selenium in relation to agriculture, p. 3–12, in: Agr. Handbook 200, US Dept. Agr., Washington, D.C. 1961
28. Martin, T.D.: Atomic Absorption Newslett. *14* (5), 109 (1975)
29. Martin, J.L., Schrift, A., Gerlach, M.L.: Phytochemistry *10*, 195 (1971)
30. McKown, D.M., Morris, J.S.: Selenium Analysis methodology and applications, p. 338–344, in: [D. D. Hemphill (ed.)] Trace Substances in Environmental Health – XI, Proc. Univers. Missouri's 11 th Ann. Conf. Trace Substances in Environm. Health, Columbia, Univers. Missouri, 1977
31. Muth, O.H., et al.: Science *128*, 1090 (1958)
32. Natl. Acad. Sci., Nat. Res. Council, Agric. Board, Committee on Animal Nutrition, Subcommittee on Selenium: Selenium, Washington, D.D., p. 79, 1971
33. Nat. Acad. Sci., Nat. Res. Council, Div. Med. Sci., Committee on Medical and Biological Effects of Environmn. Pollutants, Subcommittee on Selenium: Selenium, Washington, D.C.: Nat. Acad. Sci., P. 203, 1976
34. Neisheim, M.C., Scott, M.L.: J. Nutrition *65*, 601 (1958)
35. Nelson, A.A., Gitzhugh, O.G., Calvery, H.O.: Cancer Res. *3*, 230 (1943)
36. Nigam, S.N., McConnell, W.B.: Phytochemistry *11*, 377 (1972)
37. Nissen, P., Benson, A.A.: Biochem. Biophys. Acta *82*, 400 (1964)
38. Olson, O.E.E., Navacek, E.J., Whitehead, E.I., Palmer, I.S.: Phytochemistry *9*, 1181 (1970)
39. Pierce, R.D., Brown, H.R.: Analyt. Chem. *48*, 693 (1976)
40. Potchen, E.J.: J. Nucl. Med. *4*, 480 (1963)
41. Rook, H.L.: Analyt. Chem. *44*, 1276 (1972)
42. Schwartz, K., Foltz, C.M.: J. Amer. Chem. Soc. *79*, 3292 (1957)
43. Shrift, A., Nevyas, J., Turndorf, S.: Plant Physiol. *36*, 502 (1961)
44. Sindeeva, N.D.: Mineralogy and Types of Deposits of Selenium and Tellurium, Interscience Publ., New York, 1964
45. Smith, F.F.: Use and limitations of selenium as an insecticide, p. 41–45, in: Selenium in Agriculture, US Dep. Agriculture, Handbook No. 200, Washington, D.C., US Government Printing Office, 1961

46. Standard Methods for the Examination of Water and Wast Water, 13 th ed., 1971, Am. Pub. Hlth. Assn., Washington, p. 295–300
47. Stadtman, T.C.: Science *183*, 915 (1974)
48. Thacker, E.: The role of selenium in plants, in: Selenium in Agriculture, US Dep Agriculture Handbook No. 200, Washington, D.C., US Government Printing Office, 1961
49. Ulrich, J.M., Shrift, A.: Plant Physiol. *43*, 14 (1968)
50. Watkinson, J.H.: Analyt. Chem. *38*, 92 (1966)
51. West, P.W., Cimerman, C.: ibid. *36*, 2013 (1964)
52. Handley, R., Johnson, C.M.: ibid. *31*, 2105 (1959)
53. Nazarenko, I.I., Ermakov, A.N.: Analytical Chemistry of Selenium and Tellurium, Halsted Press, New York, 1972
54. Bectine, K.K., Goldberg, E.D.: Science *173*, 233 (1971)

Vanadium

Vanadium was discovered in 1801 by Manuel del Rio who, during his investigation of lead ores, isolated a new element and called it "erythronium". Operating under erroneous assumptions del Rio concluded that he discovered only a form of chromium. Again, in 1839, while working with iron slags, the Swedish scientiest Nils G. Sefstrom re-discovered vanadium and named the element in honor of Freya Vanadis, the Norse goddess. Finally, later that year, the name issue was settled when Friedrich Wöhler concluded that vanadium and erythronium were the same; it was not until 1927 that two Englishmen, J. W. Marden and M. N. Rich, produced 99.7% pure globules of vanadium suitable for industrial use. In its ingenue stage, vanadium was lauded for its ability to strengthen and enhance the physical properties of steel.

Production, Use and Shipment

Vanadium is found in about 65 different minerals of which carnotite, roscoelite, vanadinite, and patronite are important sources. Over 90% of the vanadium produced in the world is consumed in the production of alloy steels and iron. Most of this, about 80%, is in the form of ferrovanadium which is produced by electric-arc furnaces smelting. When added to steel in small amounts vanadium has two effects: it refines the grain of the steel matix, and with carbon present it forms carbides. Table 1 gives various uses for steel containing different concentrations of vanadium. Overall vanadium steel is especially strong and hard with improved resistance to shock. In addition, vanadium foil is used as a bonding agent in cladding titanium to steel. Vanadium also has good structural strength and a low fission neutron cross section, making it useful for nuclear applications.

In the chemical industry, compounds of vanadium are used as catalysts and in the oxidation of numerous organic compounds to commercial products, some of which can be seen in Table 2. Vanadium salts are also used to some extent in photography and ceramics, principally as a dye. The glass industry uses them as coloring agents and as a means for preventing the transmission of ultraviolet rays.

Table 1. Vanadium content of constructional steels

Steel	Vanadium content %	Typical uses
C–V	0.10–0.20	Railroad equipment-shafts, arms, connecting rods, driving axles, crank pins, guides, piston rods-parts for heavy machinery, pumps, and diesel engines
Cr–V	0.05–0.50	*Low-carbon:* Carbonized gears, camshafts, piston pins, pressure vessels, tubing, stampings
		Medium-carbon: Axles, shafts, steering arms, gears, springs, arbors, spindles, shafting, cyanided gears, bolts, washers, forgings
		High-carbon: Tools, ball bearings, wearing plates
Mn–V	0.05–0.20	Heavy forgings, plates for transportation equipment and for tanks, rivets, bolts
V-spring	0.08–0.2	Heavy and light springs for transportation equipment. Heavy machinery and instruments

Table 2. Some processes using vanadium catalysts

Production of sulfuric acid
Manufacture of phthalic anhydride
Manufacture of maleic anhydride
Production of aniline black
Oxidation of cyclohexanol to adipic acid
Oxidation of ethylene to acetaldehyde
Oxidation of anthracene to anthraquinone
Oxidation of toulene or xylene to aromatic acids
Oxidation of furfural to fumaric acid
Oxidation of hydroquinone to quinone
Oxidation of butene-2 and 1,3-butadiene to maleic anhydride
Ammonolysis/oxidation of toulene, m-xylene, p-xylene, and propylene
Preparation of vinyl acetate from ethylene
Manufacture of cyclohexylamine from cyclohexanol and ammonia
Catalytic combustion of exhaust gases
Catalytic synthesis of ethylene-propylene rubber

Consumption of vanadium has increased steadily since the early 1960's due, principally, to the increased demand for vanadium-bearing high strength, low alloy (HSLA) steels. In 1976 the apparent consumption of vanadium in the US was 4,665,900 tons [9]. The total western world capacity to produce V_2O_5 and ferrovanadium is expected to increase steadily in the coming years because of planned processing plants and the possibility to market the vanadium that can be recovered during the fuel oil refining process. The use of vanadium-based alloys is expected to increase. Currently, new steels are being tested and vanadium alloys are being considered for possible use in nuclear reactor systems.

Chemistry

Natural vanadium is composed of two isotopes, stable vanadium 51 (99.76%) and weakly radioactive vanadium-50 (0.24%), having a halflife of 6×10^{15} years. Seven

Table 3. Physical properties of vanadium [10]

Atomic number	23
Atomic weight	50.95
Crystal structure	Body-centred cubic
Lattice constant	3.034-Å
Density	6.11 g/c^3
Melting point	1,900° ± 25 °C (3,452 °F) [a]
	1,919° ± 2 °C (3,486 °F)
Boiling point	3,000 °C (5,432 °F)
Specific heat	0.12 cal/g/°C
Thermal neutron-absorption cross-section	4.98 ± 0.02 barns
Thermal conductivity	0.074 cal/cm^3/cm/s/°C at 100 °C
Electrical resistivity	24.8–26.0 microhm-cm at 20 °C

[a] Some tables give 1.710 °C ± 10 °C, based on early measurements of impure vanadium carrying carbon

other artificial, unstable isotopes have been created. Vanadium is a member of the fifth group of elements in the periodic system and a list of its physical properties is given in Table 3. Vanadium is one of the least volatile elemental metals compared with others at their melting points. Sodium hydroxide, hydrochloric acid, and dilute sulfuric acid do not dissolve vanadium. It does not tarnish in air but when heated combines readily with oxygen, nitrogen, carbon, and sulfur. Above 66 °C, the metal undergoes rapid oxidation. The four oxides corresponding to the four oxidation states are VO, V_2O_3, VO_2, and V_2O_5. The hydrogen-oxygen compounds of vanadium in the two lower oxidation states are basic and in the two higher they are partly acidic, partly basic. In acid solution the ions exhibit a lavender to blue color in the +2 and +4 states and a greenish-yellow color in the +5 state. This change of color exhibited by vanadium compounds in different valence states has allowed vanadium to be distinguished by spectrophotometric methods. Pure vanadium is a bright white metal that is soft and ductile.

Analytical Methods

Detection of vanadium in the environment has been accomplished through several analytical techniques, each displaying certain advantages and disadvantages. One proven procedure is neutron activation analysis. It has high sensitivity, can be corrected for interfering elements through instrumental or chemical means, and has the added feature of being usable on an on-stream basis. Some negative aspects are the regulations and training requirements involved with the use of radiation and the cost incurred for highly sensitive results. Practical application of this process has been in detecting vanadium levels in food, water, crude oil and biological specimens. Another process, x-ray fluorescence, is widely used in measuring vanadium concentration above 0.01%. Overall, little sample preparation is needed but in some cases of low level detection, samples must be processed to concentrate the element. This method is appropriate in analyzing fuel oils, biological materials and particulate matter filtered from the air. A third technique is spark source mass spectrometry. Although it is capable of sensitive determination of all elements simultaneously, spark source mass spectrometry is relatively uncommon, slow, costly, and complicated. It has been used, however, in analyzing airborn particles and biological samples. The fourth means of detecting vanadium is the well-known, highly accurate, and rapid atomic absorption method. It is used to analyze similar items as above and its only disadvantage is that it measures only one element at a time. Finally, atomic emission is another simple and inexpensive means of measuring vanadium concentrations in air and biological materials. The difficulty with this procedure is that the flame emission has spectral interferences that affect precision and accuracy. A summary of the state of the art, for the quantitative measurement of vanadium and metals in general, appears in Table 4.

Transport Behavior

Vanadium is widely distributed throughout nature and has been estimated at 0.017% of the lithosphere, ranking before copper, zinc, and lead. Table 5 shows the average vanadium content in various crustal materials. From these data, it should be observed that vanadium concentrations are lower in acidic igneous rocks than in basic ones and that sedimentary rocks tend to accumulate a notable amount of the element. Vanadium present in the earth's crust is usually in the trivalent state, occurring as relatively insoluble salts. Since vanadium is present in rocks and soils, it is not surprising that it is also present in water. In water, vanadium occurs in the pentavalent state, in which compounds are most soluble. The concentration of vanadium in seawater is about 2–29 µg/l and the total amount in the oceans is about 7.5×10^{12} kg [6]. Reported concentrations of vanadium in United States drinking water range from 0 to 220 ppb [11].

Accumulation and Metabolism

In some plants and animals, vanadium can reach high concentrations. The vanadium content of marine plants and invertebrate animals usually ranks above that of land plants, insects and vertebrates. An investigation done by Bertrand revealed that a species of mushroom, *Amanita muscaria*, contained more than 100 ppm of the element [7]. The biologic role of vanadium played was not deciphered but Bertrand did note that vanadium did not seem to be derived from the soil in which the mushroom grew. Table 6 shows the vanadium content in some investigated plants and animals.

Other studies investigated marine animals such as ascidia (subtype of "Tunicata") and holothurians (type "Echinodermata"). Peterson [8] is only one of the many authors that have pointed out the high vanadium concentrations common in ascidians [7]. These animals contain 10,000 times the vanadium that is present in the water in which they live [5]. In ascidians, special green blood cells called vanadocytes concentrate the element. Vanadium in these cells is in the trivalent state, is complexed to pyrrole rings, and is associated with unusually high concentrations of sulfuric acid [4]. In other ascidians, vanadium is concentrated in the blood plasma (hemovanadium) rather than in individual cells. Although the vana-

Table 4. Comparison of instrumental methods [1]

Characteristic	Neutron activation analysis	X-Ray fluorescence	Emission spectroscopy	Flame emission	Atomic absorption	Spark source mass spectrometry
Scope	Most elements	Elements heavier than Si	Most metals	Many metals	Most metals	Most elements
Sensitivity	ppb, element and neutron flux dependent	ppm, element and matrix dependent	ppm, metal and matrix dependent	ppb for few metals, ppm for many	ppb for few metals, ppm for most	ppb for most elements
Accuracy	Good	Good	Good-poor	Good-poor	Good	Good-poor
Specificity	Good	Good	Good	Good-poor	Good	Good
Contamination	Good	Good	Good	Poor	Poor	Good
Analysis time	Fast-slow	Fast	Fast	Fast	Fast	Slow

Table 5. Estimated average vanadium content of material in earth's crust (ppm) [10]. (From E. R. Rose, 1973)

Rock material	Turekian and Wedepohl (1961)	Rankama and Sahama (1950)	Kraus-kopf (1955)	Vino-gradov (1956)	Green (1959)	Champlin and Dunning (1960)	Gold-schmidt (1954)
Igneous rocks		150–315	150		90		150
Ultrabasic	40	17		140	140		
Basic	250	56–320		200	210		
Acidic	30– 88	17– 34		100	95		
Sedimentary							
Clay	120	120					
Shale	130	120	50–300	130	90–130		
Carbonate	20	Less than 10	2– 20		2– 20		
Sandstone	20	20	10– 60		10– 60		
Iron sand	600						
Green sand	68–220						
Oolithic	500						
Bog iron		Less than 10					
Laterite-bauxite	400						
Phosphorite		Less than 10					
Crude oil					30		
Asphaltic oil						23–280	
Asphalt		5,400					
Bituminous phosphates		1,600					
Coal		900				280	
Metamorphic							
Orthogneiss		5					
Paragneiss		20–70					
Quartzite		5–34					
Schist		34–56			93		
Carbonate		1.7					
Graphite		200					
Ocean		0.0003			0.001		
Deep sea sediments					330		
Upper Lithosphere							150
Meteorites		6–50					

dium complex does not seem to play a part in oxygen transport, it may be involved with the oxidation-reduction reactions that take place in the blood [11].

Two schools try to explain the source of vanadium that is incorporated into the organism. Some authors believe that organisms may obtain vanadium from marine muds while others think that organisms can concentrate vanadium from its very dilute presence in sea water. A study by Archangelskii and Kopchenova proposed a method for the systematic accumulation of vanadium in sediments and sedimentary rocks [5]. During the process of weathering, vanadium compounds are liberated from igneous rocks and transported in natural waters. Living organisms ex-

Table 6. Vanadium in plants and animals [7]

Source	Vanadium concentration (dry wt), µg/g
Plants	
Plankton	5
Brown algae	2
Bryophytes	2.3
Ferns	0.13
Gymnosperms	0.69
Angiosperms	1.6
Bacteria	+
Fungi	0.67
Animals	
Coelenterates	2.3
Annelids	1.2
Mollusks	0.7
Echinoderms	1.9
Crustaceans	0.4
Insects	0.15
Fish	0.14
Mammals	Less than 0.4

tract the vanadium from these waters and then upon their death vanadium is transferred to sediments along with the remains of the organisms.

Exposure and Toxicity

Some vanadium is present in almost all coal (consumed in the United States) [7]. Table 7 is a compilation from several sources of the estimated emission of vanadium into the ambient air resulting from coals burned in the United States for the year 1969.

In addition to its presence in coal, all crude oils of petroleum origin contain some vanadium. Its presence here is represented by several type of vanadium complexes (as seen in Fig. 1), the most common being the porphyrins. When crude oil is processed and refined for its commercial use almost all the vanadium is left in the residue. This is the heavy end of the distillation process and concentrations of vanadium ranging from 1 to 60,000 ppm are common [3]. The vanadium content of residual fuel oils, however, varies widely depending on such factors as the origin of the crude oil from which they are prepared, the amount of residual fuel that is extracted from the crude oil, and the extent to which the residual fuel was processed to achieve an acceptable sulfur level. The burning of these residual fuels is the source of much of the vanadium that is released to the atmosphere. Measurements in Osaka/Japan of the average vanadium concentration in the atmosphere, related to the quantity of fuel oils burned in a 2 km · 2 km area, range from 0.05 to 0.55 µg/ m^3 [12]. In cases where residual fuel oils were processed in order to reduce their sulfur content, the vanadium content was also seen to decline considerably. What

Table 7. Emission of vanadium to the ambient air by burned coal [7]

Type and use of coal	Coal, 1,000 tons	Vanadium in coal, tons	Vanadium in fly ash, tons	Control of fly ash, %	Vanadium emitted to air, tons
Bituminous					
Electric-power utilities	308,462	9,254	6,015	85	902
Manufacturing	93,248	2,797	1,818	60	727
Retail deliveries	12,665	380	247	50	124
Coking	92,901	2,787	–	100	0
Subtotal	507,276	15,218	8,080		1,753
Anthracite	9,275	1,159	753	50	377
Total					2,130

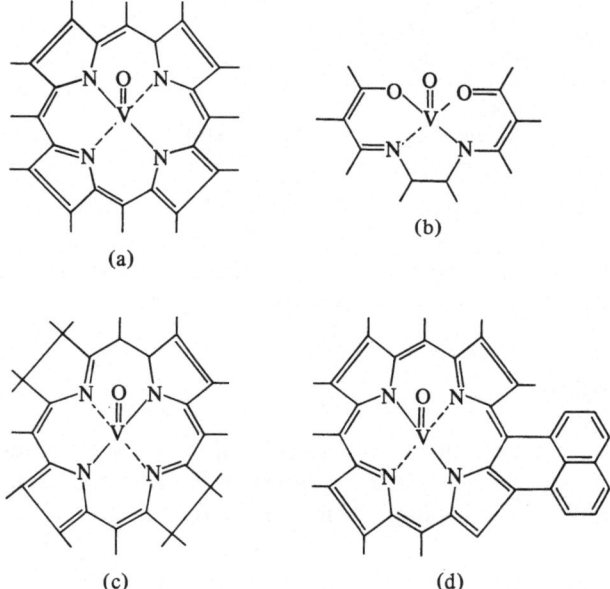

Fig. 1a–d. Key types of vanadium complexes in petroleum: **a** porphyrins, **b** mixed ligands such as N_2O_2 type, the B-ketimines, **c** pseudoaromatic pheophorbides such as a bacterio-chlorophyll (the outer conjugation is interrupted, but still belong to the diaza-18-annulene system), and **d** highly aromatic porphins such as the dehydrogenated product of m-α-naphthylporphyrin, which was identified in Nonesuch shale. Type **a** is found in all petroleum. Types **b**, **c**, and **d** are commonly referred to as the non-porphyrin type of vanadium [13]

is lacking in studies concerning the combustion of residual fuel oils is information regarding the chemical form in which vanadium is emitted into the air. Finally, in contrast to the heavy fuel oils and crude oils, petroleum that has been distilled (automotive diesel fuels, motor and aviation gasolines, home heating oils) contains little or barely detectable traces of vanadium (0.05 ppm or less). Therefore, the burning of these fuels contributes hardly any vanadium to the atmosphere.

Table 8. Uptake of vanadium (^{48}V) in some vegetables [a] [11]

	Ash wt %	Uptake of ^{48}V	
		Dry wt %	Wet wt %
Lettuce	0.211×10^{-4}	1.01×10^{-5}	0.68×10^{-6}
Spinach	0.595×10^{-4}	2.09×10^{-5}	1.72×10^{-6}
Dill	1.05×10^{-4}	3.16×10^{-5}	2.98×10^{-6}
Parsley	5.66×10^{-3}	1.50×10^{-3}	1.45×10^{-4}
Carrots, peeled	6.55×10^{-4}	6.02×10^{-5}	0.91×10^{-5}
Radishes, peeled	1.58×10^{-3}	2.53×10^{-4}	0.21×10^{-4}
Potatoes, peeled	7.87×10^{-4}	2.27×10^{-4}	0.34×10^{-4}
Radishes, leaves	6.28×10^{-4}	1.49×10^{-4}	0.14×10^{-4}
Potatoes, leaves	2.29×10^{-4}	4.41×10^{-5}	0.30×10^{-5}

[a] The indicated uptake is here as % ^{48}V (of the entire amount of ^{48}V in the soil of the box where the vegetable grew) per gram of the vegetable

Table 9. Lethal doses of vanadium compounds [2]

Compound	Lethal dose, mg/kg			
	Rabbit	Guinea pig	Rat	Mouse
Colloidal vanadium pentoxide	1 – 2	20–28		87.5–117.5
Ammonium metavanadate	1.5– 2.0	1– 2	20– 30	25 – 30
Sodium orthovanadate	2 – 3	1– 2	50– 60	50 –100
Sodium pyrovanadate	3 – 4	1– 2	40– 50	50 –100
Sodium tetravanadate	6 – 8	18–20	30– 40	25 – 50
Sodium hexavanadate	30 –40	40–50	40– 50	100 –150
Vanadyl sulfate	18 –20	35–45	158–190	125 –150
Sodium vanadate		30–40	10– 20	100 –150

Little is known on the vanadium content in foods and only recently have some studies been attempted. Using the technique of neutron activation analysis, Soremark [11] studied the vanadium concentration of some vegetables. His results are reproduced in Table 8. Parsley, radishes, dill and lettuce showed the highest concentrations. Soremark concluded that although the presence of vanadium in foods is uneven, there does seem to be a rather regular dietary intake.

Experiments designed to test vanadium toxicity on experimental animals have shown that when administered orally the toxicity is high, when administered parentally. low, and intermediate by the respiratory tract. Table 9 gives the lethal doses of various vanadium compounds given by injection to 4 types of experimental animals [2]. In their studies utilizing rabbits exposed to pure vanadium pentoxide dust at 205 mg/m^3, Faulkner-Hudson noted that the rabbits displayed acute tracheitis and broncho-pneumonia with pulmonary edema. After ashing the lungs, liver, kidney, and intestines, vanadium was seen to be present in small quantities.

In similar but long-term investigations conducted on rabbits, the results were comparable to the above study except that no vanadium was detected in the intestine.

Vanadium has been shown to have many metabolic effects. One study showed, that rats fed vanadium pentoxide had a lowered cystine content in their hair, and workers exposed to vanadium a lowered cystine content in their fingernails [7]. This indicates that vanadium causes a decrease in the synthesis of cystine and cysteine. More research on the toxicity of vanadium and its salts is needed.

Various studies have been undertaken which have added to the information concerning the biological effects of vanadium on man. Industrial exposure has provided a prime avenue for investigation. The most common method of entry of vanadium, due to industrial exposure, is via the respiratory system, although entry by the gastro-intestinal system is also possible. The first reliable information pertaining to the effects of vanadium dust in man is cited by Symanski [7]. Based on his observations of 19 individuals who had been in contact with vanadium for extended periods of time, he concluded that their ailments were solely due to the irritation effects of vanadium on the mucous membranes of the respiratory tract and on the conjunctiva. "In all cases, he observed conjunctivitis with smarting eyes, rhinitis with watery discharge, sore throat, persistent cough with pressure over the chest, possibly a stitch, and many rhonci" [7]. In addition to these findings, Symanski stated that there were no gastrointestinal symptoms, no indications of lethargy of the blood-forming organs, nor any symptoms of organic injuries of the kidneys or the nervous system.

In another study, concerning vanadium intoxication suffered by boiler cleaners, Williams reported some additional symptoms [7]. "The primary symptoms occurred 0.5–12 h after starting work and consisted of a dry cough, sneezing, severe dyspnea, lassitude, and depression, with a disinclination to follow the usual evening activities" [7]. An esthetic symptom also caused while in contact with the petroleum soot, was a greenish-black coating on the workers tongues. This coloration which resulted from the deposition of vanadium slats of reduced-valence forms, was found to be harmless. The results of this investigation stated in all cases, the symptoms were only temporary and cleared up after removal from exposure to the vanadium containing dust.

Experimental studies of vanadium using human subjects have been scarce but a few have been accomplished. One such investigation involved 9 healthy subjects, aged 27 to 44 years [7]. In one test, two volunteers were accidently exposed for 8 h to 1 mg/m^3 of vanadium pentoxide dust instead of the planned 0.05 mg/m^3. After 5 h some intermittent coughing developed that was deemed psychological in nature. By the end of the test, though a persistent cough had commenced which lasted for a period of 8 days. Other than a cough, there were no other signs of aberrations in any bodily functions. Three weeks after the test, the same volunteers, were inadvertently engulfed in a heavy cloud of vanadium pentoxide dust for a period of 5 min. Symptoms from this exposure ranged from coughing up sputum to rales and expiratory wheezings.

Overall, these tests support the assumptions that exposure to vanadium dust causes irritating effects in the mucous membranes and the respiratory tract.

Vanadium salts are not readily absorbed in the intestine. They are mainly excreted in the urine and to a lesser degree in the stool. Even when highly soluble forms of vanadium are induced into the body only about 1% is absorbed and most of this is then excreted rapidly. This low absorption and high excretion rate tends to make vanadium less toxic than other metals that do not possess these characteristics. "Roschin pointed out that the degree of vanadium toxicity depends on the dispersion and solubility of vanadium aerosols in biologic media" [7]. It was also noted that the toxicity depends on valence, that is, it increases with increasing valence, with pentavalent vanadium being the most toxic [7].

In the United States, The American Conference of Governmental Industrial Hygienists have proposed threshold limit values of 0.5 mg/m^3 for vanadium pentoxide dust or vanadates and 0.105 mg/m^3 for vanadium pentoxide fume. These limits are based on information compiled from industrial experience and from experimental human and animal studies. The Russians have proposed a more stringent limit, 0.1 mg/m^3, for dust concerning their work with rabbits.

References

1. Braier, H.A., Eppolito, J.: Determination of Trace Metals in Petroleum Instrumental Methods, in: The Role of Trace Metals in Petroleum [Yen, T.F. (ed.)], p. 64–85, Ann Arbor, Mich., 1975
2. Faulkner-Hudson, T.G.: Vanadium Toxicology and Biological Significance, p. 140, New York, Elsevier, 1964
3. Fischer, R.P.: American Oil Refineries: Can they yield Vanadium? in: The Role of Trace Metals in Petroleum [Yen, T.F. (ed.)], p. 200–205, Ann Arbor, Mich. 1975
4. Levine, E.P.: Science *133*, 1352 (1961)
5. Manskaya, S.M., Drozdova, T.V.: Geochemistry of Organic Substances, p. 347, Pergamon Press, New York, 1968
6. Mason, B.H.: Principles of Geochemistry (3rd ed.), p. 329, New York, John Wiley, 1966
7. Natl. Res. Council, Committee on Biolog. Effects of Atmospheric Pollutants, Vanadium, Washington, D.C., Nat. Aca. Sci., p. 117, 104 (1977)
8. Peterson, P.J.: Sci. Prog. *59*, 505 (1971)
9. Rausch, B.A.: Eng. Mining J. *178*, 104 (1977)
10. Rose, E.R.: Geology of Vanadium and Vanadiferous Occurrences of Canada, Economic Geology Rep. No. 27, p. 130, Dep. Energy, Mines and Resources, Canada, 1973
11. Soremark, R.: J. Nutr. *92*, 183 (1967)
12. Sugimea, A., Hasegawa, T.: Env. Science and Technol. *7*, N5, 444 (5) (1973)
13. Yen, T.F. (ed.): The Role of Trace Metals in Petroleum, p. 221, Ann Arbor Sci. Publ. Inc., Ann Arbor, Mich., 1975

Acknowledgement

The author wishes to thank the TCU Research Foundation for its support during the writing of this chapter and to Cindy Sander for her assistance in preparation of the final manuscript.

C$_1$ und C$_2$ Halocarbons

C. R. Pearson

Imperial Chemical Industries Ltd., Brixham Laboratory
Overgang Brixham, Devon TQ5 8BA, England

Introduction

Public concern over the possible effects of chemicals on man and his environment has largely focussed on a small number of classes of compounds; of these, chlorinated hydrocarbons are perhaps the best known. A cursory examination of the scientific literature could easily lead to the impression that they consist entirely of chlorinated pesticides and polychlorinated biphenyls; but in terms of total quantity produced, these are less important than many others. This Chapter discusses another class, the aliphatic C$_1$ and C$_2$ Halocarbons, which includes most of the halogenated products of the Organic Chemicals industry, which are of major commercial importance. (Table 1); it excludes all those containing Fluorine.

Other significant classes which are discussed in separate chapters are Chlorinated Aromatics, Fluorocarbons and Chloroaromatic Compounds containing oxygen. The production processes for all the products mentioned in this chapter are closely related, and often integrated on a common site; but initially they will be considered as 3 arbitrarily chosen groups, namely Chloromethanes, Halogenated C$_2$ compounds (ethanes and ethenes) and Vinylchloride/Vinylidenechloride.

The environmental behaviour and possible effects of individual compounds are related to their use and properties and bear little relation to the arbitrary production groups; in all further discussion, therefore, the class will be considered as a whole.

Production Processes

There are 3 basic routes for the synthesis of all the compounds in this class; all involve the introduction of chlorine into a short chain aliphatic hydrocarbon feedstock [2, 14, 15, 19, 20, 50].

Table 1. Production and uses of halogenated aliphatic hydrocarbons of commercial importance

Name	Synonyms	World production million tonnes/ year	Major uses
Chloromethane	Methyl chloride	0.4	Intermediate
Dichloromethane	Methylene dichloride	0.5	Solvent
Trichloromethane	Chloroform	0.25,	Intermediate, solvent
Tetrachloromethane	Carbon tetrachloride	1.0	Intermediate Fumigant
Bromomethane	Methyl bromide	0.02	Fumigant
Chloroethene	Vinyl chloride	10.00	Monomer for plastics intermediate
1,1-Dichloroethene (asym)	Vinylidene chloride	0.2	Monomer for plastics
Trichloroethene	Trichloroethylene	0.6	Solvent
Tetrachloroethene	Perchloroethylene	1.1	Solvent
Chloroethane	Ethyl chloride	0.4	Solvent intermediate
1,2-Dichloroethane	Ethylene dichloride	13.0	Intermediate
1,1-Dichloroethane	Ethylidene chloride	(0.5)	Intermediate
1,1,1-Trichloroethane	α-tri Methyl chloroform	0.6	Solvent
Hexachloroethane		No data	Various small
Bromo ethane	Ethyl bromide	< 0.1	Fumigant
1,2-Dibromoethane	Ethylene dibromide	0.25	Antiknock scavenger
3-Chloropropene-1	Allyl chloride	0.5	Intermediate
3-Chloropropane, 1–2 epoxide	Epichlorhydrin	0.5	Monomer for resin
2-Chlorobutadiene	Chloroprene	0.3	Monomer for synthetic rubbers

(a) Direct chlorination, either in liquid or vapour phase; the former tend to be addition reactions, the latter substitution,

e.g: $$Cl_2 + C_2H_4 \rightarrow C_2H_4Cl_2$$

$$2Cl_2 + CH_4 \rightarrow CH_2Cl_2 + 2HCl.$$

(b) Hydrochlorination

e.g: $$HCl + C_2H_4 \rightarrow C_2H_5Cl.$$

The reverse form of this, dehydrochlorination or thermal cracking, will normally be carried out at higher temperatures and pressures.

e.g: $$C_2H_4Cl_2 \rightarrow CH_2 = CHCl + HCl.$$

(c) Oxychlorination in which air is added under controlled conditions,

e.g: $$2HCl + C_2H_4 + \tfrac{1}{2}O_2 \rightarrow C_2H_4Cl_2 + H_2O.$$

The most important hydrocarbon feedstocks are methane and ethylene; acetylene and ethane are also significant, and propylene, propane and various waste streams may also be used.

The attraction of integrated processes lies both in the ability to reuse by-product HCl, and to vary the balance of products to match demand.

Most of the other by-products of the reactions are themselves chlorinated; taking reactions to completion tends to produce the per-compounds, containing 1, 2 or more carbon atoms, such as carbon tetrachloride, hexachloroethane and hexachlorobutadiene.

The important production processes for the major groups of compounds are outlined below; details will vary from plant to plant, and because of their commercial value, will not normally be publicly available.

Halomethanes

All 4 chlorocompounds, namely methyl chloride, methylene dichloride, chloroform and carbon tetrachloride, are of commercial importance; they are produced by successive chlorination of methane, followed by separation and purification of the halomethanes mixture [40].

An alternative feedstock of growing significance is methanol, which may be transported much more easily than methane. In this case, the first stage is hydrochlorination which yields methyl chloride, which is then further chlorinated.

A traditional process for the production of carbon tetrachloride, which is still in use, consists of the reaction of chlorine with Carbon Disulphide; conversion efficiency is high and there are few chlorinated by-products. Sulphur is recovered as such, or reconverted to CS_2.

Carbon tetrachloride may also be co-produced with perchloroethylene, by chlorination of propane, propylene, ethylene, or on its own from a whole series of chlorinated residues.

Methyl bromide is produced by hydrobromination of methanol

$$CH_3OH + HBr \longrightarrow CH_3Br + H_2O.$$

Other halomethanes may be synthesised by routes analogous to the chloromethanes, but are not of commercial significance.

Halogenated C₂ Compounds

The major products in this group consist of trichloroethylene, perchloroethylene, and 1,1,1-trichloroethane; ethylene dichloride is strictly speaking within the group but as its major use is in the production of vinyl chloride, the two will be considered together in the next section.

The early routes to both tri- and perchloroethylene used acetylene as a raw material; chlorination was followed by dehydrochlorination. The preferred raw material is now ethylene, proceeding via ethylene dichloride.

1.1.1-trichlorethane is normally produced from vinyl chloride, by hydrochlorination to 1,1-dichloroethane, followed by further chlorination.

Ethyl chloride, ethyl bromide and ethylene dibromide are similarly prepared by halogenation of ethane, or via hydro-halogenation of ethanol.

Vinyl Chloride and Vinylidene Chloride

The older route to vinyl chloride used acetylene as raw material, in a one step hydrochlorination

$$C_2H_2 + HCl \longrightarrow CH_2 = CHCl.$$

Modern processes use ethylene as raw material; the first step consists in the production of ethylene dichloride by chlorination, to be followed by a dehydrochlorination. Ethylene dichloride thus occupies a key position as the major raw material for the production of all the chlorinated ethylenes.

Vinylidene chloride is produced by dehydrochlorination of trichloroethane (1,1,1- or 1,1,2-).

Other Compounds

Allyl chloride is produced by direct chlorination of propylene; further reaction with chlorine and water, by chlorohydrination and hydrolysis, yields epichlorhydrin.

Small quantities of hexachloroethane are produced by direct chlorination of ethane. Chlorobutadienes may be obtained by chlorination of butadiene; hexachlorobutadiene occurs as a by-product in the heavy ends of many of the processes described above, from which it may be isolated.

Physical Properties

The members of this class are typically mobile liquids except for methyl chloride and vinyl chloride, which are gases at normal temperatures; and hexachloroethane, which is a solid.

Some relevant physical properties are shown in Table 2. In any series formed by successive chlorination of a hydrocarbon molecule, the properties normally vary directly with the amount of chlorine present in the compound. The percentage of chlorine by weight in most of the compounds listed is higher than that in typical PCB and chlorinated insecticides; by contrast, they are very much more volatile, and more soluble in water.

Methods of Analysis

The most widely used method of detection and quantification of halocarbons is Gas Chromatography, normally with an electron capture detector [12, 40]. In some cases, air or water samples can be injected directly into the instrument; where initial levels are low, a concentration step consisting either of adsorption onto active carbon, or solvent extraction, may be necessary. In the case of very volatile compounds like vinyl chloride or methyl chloride, special techniques are needed to prevent loss from the sample, and to obtain good resolution.

Table 2. Physical properties of some aliphatic halocarbons

Trivial name	Formula	Molecular weight	% Chlorine	% Bromine	No.Cl atoms	M.P. °C	B.P. °C	Vapour pressure mm Hg @ 20 °C	Solubility [36] in water mg/l @ 20 °C	Partition [36] coefficient water/air 20 °C w/v per w/v
Methyl chloride	CH$_3$Cl	50	70	–	1	– 98	– 24	3,756	7,250	3.3
Methylene chloride	CH$_2$Cl$_2$	85	84	–	2	– 95	40	362	13,200	8.1
Chloroform	CHCl$_3$	119	88	–	3	– 64	62	151	8,200	8.6
Carbon tetrachloride	CCl$_4$	154	92	–	4	– 23	77	90	785	1.1
Methyl bromide	CH$_3$Br	95	–	84	–	– 93	5	1,420	900	1.1
Vinyl chloride	CH$_2$CHCl	63	56	–	1	– 153	– 14	2,320	60	0.02
Vinylidene chloride	CH$_2$CCl$_2$	97	73	–	2	– 122	32	497	400	0.16
Dichloroethylene	CHCl-CHCl	97	73	–	2	– 80	60			
Trichloroethylene	CHClCCl$_2$	131	81	–	3	– 73	87	58	1,100	2.74
Perchloroethylene	C$_2$Cl$_4$	166	86	–	4	– 19	121	14	150	1.22
Ethyl chloride	C$_2$H$_5$Cl	65	55	–	1	– 136	12	1,000	5,740	
Ethylene dichloride	CH$_2$ClCH$_2$Cl	99	72	–	2	– 35	84	64	8,800	26.4
Methyl chloroform	CH$_3$CCl$_3$	133	79	–	3	– 30	74	96	480	0.71
Tetrachloroethane	C$_2$H$_2$Cl$_4$ (sym)	168	84	–	4	– 43	146	5	2,900	6.3
Pentachloroethane	C$_2$HCl$_5$	202	88	–	5	– 29	160	3.4	340	11.8
Hexachloroethane	C$_2$Cl$_6$	237	90	–	6	187	187 (s)	0.4	50	
Ethyl bromide	C$_2$H$_5$Br	109	–	73	–	– 199	38	380	9,100	
Ethylene dibromide	CH$_2$BrCH$_2$Br	188	–	85	–	– 10	131	11	4,000	
Allyl chloride	CH$_2$ClCHCH$_2$	77	46	–	1	– 134	45	340	100	
Chloroprene	CH$_2$CHClCH$_2$	89	40	–	1	– 130	59	190	–	
Propylene dichloride	CH$_2$ClCHClCH$_3$	113	63	–	4	– 100	97	0.42	2,700	
Hexachlorobutadiene	C$_4$Cl$_6$	261	82	–	6	– 20	215	0.15	2	0.97

As in any GC process, identification relies on comparing retention time on the column with that of a standard; use of the Electron Capture detector, which is particularly sensitive to halogens, gives a greater measure of confidence to this process. But because of this high sensitivity, there is a wide range of response to different chlorinated compounds depending on the number of chlorine atoms in the molecule; there is a marked fall in sensitivity below about 80% chlorine content, and in some situations it may be better to use the more rugged Flame Ionisation detector.

Where analysis of sediments, foods or biological tissues is required, it is necessary to carry out a solvent extraction, followed by clean-up of the extract to remove non halogenated fatty material. Great care in operation, and the use of highly purified solvents, are needed, as many of the compounds under discussion are widely used in and around laboratories, or in the construction and maintenance of apparatus and equipment.

It is possible to include all the compounds discussed in this chapter on a single GC trace [5, 47, 54], but for greater sensitivity and resolution, the low and high boilers will be analysed separately. Such techniques allow large quantities of data to be obtained on a routine basis, at comparatively low cost.

Greater certainty in the identification of compounds, particularly from unknown sources, may be obtained by the use of a Mass Spectrometer, used in series with Gas Chromatography [48]. It has a further advantage in being about equally sensitive to all compounds, regardless of the chlorine content. While GC/EC ranges in sensitivity from about 1 part in 10^6 for ethylene dichloride, to 1 part in 10^{12} for carbon tetrachloride, GC/MS will give a routine sensitivity of 1 part in 10^9 for all compounds.

Some efforts have been made to use long path infra-red spectroscopy, either with an artificial source, or the sun itself, to make real-time measurements in the atmosphere [23, 46]. But for various reasons, its applicability is limited.

Some of the higher molecular weight, higher boiling halogenated impurities and by-products are not volatile under GC conditions; these may be determined by High Performance Liquid Chromatography [29].

There is a growing interest in the quantification of total organic halogens in environmental samples. Traditional estimates have always been made by adding up all the individual identified components; doubt is being cast on this procedure by the growth in the number of previously unsuspected compounds which are now being identified, especially those of natural origin. The best known method involves pyrolysis followed by micro-coulometry, but neutron activation techniques may also be employed [1, 22].

Use Patterns and Losses to the Environment

Estimates of the total world production of the major aliphatic halocarbons, together with their principal uses, are shown in Table 1. It is assumed that the major producers are USA, Western Europe and Japan (no figures are available for Eastern Europe or China); and when only US figures are available, that they represent

50% of the total world consumption [1, 19, 20, 26, 39, 50]. Figures are rounded off to typical 1978–1980 levels.

Annual production figures do not, of course, represent the inputs to the environment; this will only be the case when the compound in question is used dispersively, or completely rejected after a short period of closed use. When a compound is used as an intermediate for further processing, only the small percentage lost during production will enter the environment. There are other significant inputs to the environment, in addition to the major commodity chemicals themselves, both man made and of complete or partial natural origin. These include wastes and by-products from manufacture of the major chemicals; the reaction products of chlorine with organic matter in water and effluent treatment; and compounds generated in the environment itself.

Products which are Largely Discharged after Use

These will consist of those materials which are used principally as solvents or fumigants, which include the majority of the methylene chloride, methyl bromide, trichloroethylene, perchloroethylene, methyl chloroform, ethyl bromide and a small proportion of the chloroform, carbon tetrachloride, ethyl chloride, ethylene dichloride and ethylene dibromide.

Losses During Manufacture and Processing of Primary Products

Even in a well operated, integrated production unit some losses of both original reactants and products will always occur. Such inputs are significant in the case of those compounds whose major use is as intermediates for further synthesis; these include most of the methyl chloride, chloroform, carbon tetrachloride, vinyl chloride, vinylidene chloride, ethyl chloride, ethylene dichloride, allyl chloride, epichlorhydrin and chloroprene. Estimates that have been made suggest that losses of between 0.5% and 2% of raw material and finished product occur, depending on the age of the plant and the nature of the process [1, 14, 15].

By-product Formation

A characteristic of many modern integrated productions units is their flexibility in terms of balance of products, which are separated from each other, and purified, by distillation. The "lights" fraction from one product will therefore consist largely of a mixture of several other products in the class, and so need not be separately considered. The "heavies" fractions contain a number of highly chlorinated aliphatic compounds of chain lengths C_2–C_4, and higher, of which the most common and characteristic is hexachlorobutadiene (HCBD), and also hexachlorobenzene (HCB). Estimates suggest a world output of 10,000 tons of HCBD in the heavy ends of all the chlorination processes, with lesser quantities of pentachlorobutadiene, and penta and hexachlorethane.

One other major by-product is that arising from the manufacture of propylene oxide, using the old chlorhydrin route; hydrolysis of the latter yields a mixture consisting of propylene dichloride (about 75%), dichlorodi-isopropylether (20%), some chloroacetone and other minor constituents [4]. A total input of 50,000 tpa propylene dichloride was probably the maximum, which will now be falling as direct oxidation synthetic routes become available.

Products of Chlorination of Water and Effluents

The use of chlorine for disinfection of potable water has long been practised, it has recently been found [52, 54] that this can lead to a significant increase in the level of Trihalomethanes (THM). The important THM are chloroform, bromodichloromethane, chlorodibromomethane, and bromoform; the levels detected range from 1 µg/l to 100 µg/l and above as total THM, of which typically 80% will be chloroform. THM arise from the chlorination of a number of precursors, many of them humic in nature. The bromine is believed to arise from bromide present in the water, which liberates bromine in the presence of excess chlorine.

A number of other higher molecular weight chlorinated materials is also produced, especially when treated sewage or effluents are chlorinated [22, 29], or when chlorine is used in bleaching of textiles or paper pulp.

Some estimate of the chloroform added to the environment by chlorination of drinking water may be made as followes:

For a population of $1,000 \times 10^6$, with a water use of 100 l/day each, at a mean $CHCl_3$ level of 20 µg/l, the annual production of chloroform is ca 7,000 tons.

Formation in the Environment

Disposal of waste from human activities will involve all the materials mentioned in the previous sections; it will also include quantities of chlorine containing polymers such as PVC, which will be either tipped on land or burnt. PVC is a polymer of low reactivity, but there are no data on its ultimate chemical stability after soil burial, or on any identified products of decomposition.

Combustion in incinerators, or by open burning, may lead to the production of low molecular weight oligomers, or of chlorinated C_1–C_2 materials such as methyl chloride. Methyl chloride has been postulated as a major product of the combustion of agricultural waste, and of slash and burn land clearance [42]; estimates suggest that as much as 5×10^6 tonnes per year of methyl chloride could be emitted from all fires throughout the world. There have been suggestions that both chloroform and carbon tetrachloride may have a natural atmospheric source, by chlorination of methane [34]; but there is no confirmation, despite their ubiquitous presence, and the known occurrence of free chlorine.

Although it has long been held that the carbon-chlorine bond is too stable for it to be formed by biological systems, there is strong evidence that large quantities of methyl chloride are produced as a result of the activities of marine algae; this may well result from interchange reactions between chloride and methyl iodide, methyl bromide or dimethyl sulfide, all of which have been associated with the metabolism of specific algae.

Estimates as high as 40,000 tonnes per annum of methyl chloride production by marine algae have been made.

Table 3. Estimated total inputs to the environment (kilotons per annum)

Compound	Lost after use	Production losses	Production by-products	Water chlor- ination	Natural origin	Total
Methyl chloride	–	4	–	–	5,000+	5,000+
Methylene chloride	500	–				500
Chloroform	5	5		10?	?	20+
Carbon tetrachloride	10	20			?	50+
Vinyl chloride		200			?	200
Vinylidene chloride		2				2
Trichloroethylene		600				600
Perchloroethylene		1,100				1,100
Ethyl chloride		15				15
Ethylene dichloride		1,200				1,200
Methyl chloroform		600				600
Hexachloroethane		5				5
Allyl chloride		5				5
Chloroprene		5				5
Propylene dichloride			50			50
Hexachlorobutadiene			10			10

Total Inputs to the Environment

A summary estimate of total annual sources of the major chlorinated aliphatics is presented in Table 3 [39, 40, 44].

Occurrence in the Environment

Data on the occurrence of low molecular weight chlorinated compounds have until recently been very sparse; but the impact of 2 issues, namely the effect of fluorocarbons on the ozone layer, and the suspicion that the trihalomethanes found in chlorinated waters, and some solvents, are carcinogenic, has led to a rapid increase in reported analysis in air, and in surface and drinking water.

The Atmosphere

Much of the data on the occurrene of chlorinated hydrocarbons in the atmosphere has been collected as part of the study of the effect of fluorocarbons on the ozone layer.

The atmosphere consists of a lower layer, the troposphere, separated from the upper stratosphere by the tropopause, a boundary characterised by a sharp reversal of rate of change of temperature. The only non-fluorinated halocarbons found in the stratosphere are carbon tetrachloride, and some methyl chloroform and methyl chloride. The troposphere, however, contains a much wider range of chlorocarbons at detectable levels.

Although there are significant variations from place to place, between urban and rural areas, between marine and continental masses, and between northern

Table 4. Typical concentrations of halocarbons in the atmosphere [35, 39]

Compound	Concentration parts $\times 10^{-9}$ by volume
Methyl chloride	0.6
Methylene chloride	0.03
Chloroform	0.03
Carbon tetrachloride	0.15
Methyl chloroform	0.10
Trichloroethylene	0.005
Perchloroethylene	0.005
Methyl bromide	0.05
Methyl iodide	0.01

and southern hemisphere, the troposphere is sufficiently well mixed for sensibly uniform concentrations of chlorocarbons to be found throughout the world; typical values are shown in Table 4.

Vinyl chloride, vinylidene chloride and ethylene dichloride are not normally detected in the atmosphere.

Some early reports of atmospheric concentrations caused confusion, by not consistently using either w/v or v/v units; this can result in what are essentially identical results appearing to differ by a factor of 10^3.

Much higher concentrations occur in indoor atmospheres, and also in urban areas where vehicles are operating; chloroform and methyl bromide are significant components in motor car exhausts. The use of chloroform, carbon tetrachloride and methyl bromide as fumigants will also lead to high concentrations both in the surrounding areas and in the treated food [40].

Surface Sea Water and Sediments

In spite of the importance of the ocean as a source or sink for many halogenated hydrocarbons, there are surprisingly few data reported on occurrence in surface water, and less in the deep oceans. Early work suffered from an inability to completely separate and identify all compounds, particularly those of low molecular weight. Representative figures are shown in Table 5, including some for inshore sediments in an area contaminated by waste discharges.

As in the atmosphere, vinyl chloride, methylene chloride and ethylene dichloride are not detected, even in inshore water. Waters from an area known to contain industrial effluents contained traces of tetrachloroethane, pentachloroethane, and pentachlorobutadiene; but even in areas where propylene oxide wastes are discharged, propylene dichloride is rarely found [43, 44].

Fresh Waters

The occurrence of halogenated compounds in rain and snow is poorly documented, but broadly parallels their distribution in the atmosphere [43, 51]; concen-

Table 5. Representative concentrations of chlorinated hydrocarbons in sea waters [34, 40, 43, 51]

Compound	Conc. in water (Parts in 10⁹ w/w)		Typical conc. on inshore sediments (Parts in 10⁹ w/w)
	Typical open ocean	High in inshore waters	
Methyl chloride	0.015	–	–
Chloroform	0.015	0.1	0.5
Carbon tetrachloride	0.005	0.15	0.5
Methyl chloroform	0.010	0.15	0.25
Trichloroethylene	0.010	0.3	0.500
Perchloroethylene	0.010	0.15	1.000
Hexachlorobutadiene	0.001	0.005	0.5
Methyl iodide	0.080	0.2	–

trations are summarised in Table 6. Methyl chloride is not reported; the concentration of chloroform is unexpectedly high, much more so than that of carbon tetrachloride; the higher values in rain are reported from an urban/industrial area. The relatively high concentration of chloroform, compared with carbon tetrachloride, in both sea and fresh surface waters could be explained by its higher water/air partition coefficient (Table 2).

There is by contrast a very large number of observations of distribution in surface and underground waters, in Germany, Sweden, Switzerland, UK and USA; in many cases, where such waters are used as a source for potable water supply, there are parallel observations on the finished drinking water after treatment [1, 7, 21, 30, 33, 43, 49, 51, 52]; these also are summarised in Table 6.

The general impression is of a widespread distribution, at the μg/l level, of chloroform, carbon tetrachloride, trichloroethylene, perchloroethylene and methyl chloroform; methyl chloride, vinyl chloride and ethylene dichloride are virtually absent, despite the large quantity of each known to be produced; while methylene chloride is reported locally, and such compounds as hexachlorobutadiene at very low levels only in waters known to be affected by production effluents. There is, nevertheless, a very wide variation about these typical values, of at least 3 orders of magnitude; it is difficult, therefore, to quote fully representative levels. Surface waters provide a much lower potential for widespread mixing than does the atmosphere, where concentrations of constituents are much more consistent. Many surface waters also contain the 3 brominated THMs; these are probably derived from discharges of chlorinated potable water after use.

Underground waters in general have much lower concentrations than rivers and lakes; but here again, there are high local concentrations [21], and they often contain a much higher proportion of brominated derivatives than surface waters.

Potable waters supplied after treatments which include chlorination contain much higher levels of trihalomethanes than the raw water they are derived from, especially of chloroform; but by contrast they contain much less carbon tetrachloride, trichlorethylene, perchloroethylene and methyl chloroform, probably as a result of chemical destruction by the chlorine.

Table 6. Typical ranges of concentrations of halocarbons found in fresh waters [7, 8, 10, 17, 21, 24, 30, 33, 43, 44, 49, 51] (Parts in 10^9 w/v)

Compound	Snow	Rain	Surface waters		Underground waters		Drinking waters		Normal limit of detection
			Typical	High urban areas	Typical	High	Typical	High	
Methylene chloride	–	–	0.05	0.5	–		–	–	0.05
Chloroform	0.002–0.12	0.005–0.2	0.3	2.0–10	0.3	5	20	100	0.01
Carbon tetrachloride	0.0003–0.003	0.001–0.3	0.1	1.5–5	0.1	2	0.1	0.5	0.001
Dichlorobromomethane	–	–	0.1	0.5	0.1	0.5	10	40	0.01
Chlorodibromomethane	–	–	0.05	0.5	0.05	0.5	5	10	0.01
Bromoform	–	–	0.05	0.1	0.05	0.5	1	40	0.01
Trichloroethylene	0.001–0.04	0.005–0.15	0.5	1.0–20	0.2	2	0.1	0.1	0.002
Perchloroethylene	0.001–0.02	0.001–0.15	0.5	1.5–20	0.2	5	0.1	0.1	0.002
Methylchloroform	0.001–0.03	0.005–0.09	0.1	1.0–10	0.2	5	0.1	0.1	0.005

Sewage and Industrial Effluents

Sewage, especially after chlorination, and industrial effluents have been recognised as a significant source of chlorinated hydrocarbons entering the aquatic environment [8, 10, 17, 33]. Chloroform, trichloroethylene, perchloroethylene and methyl chloroform are commonly found at the 10–100 µg/l level, with rather less methylene chloride, in crude sewage; these levels are reduced by 90–99% during the aerobic biological treatment stage, probably by volatilization.

In industrial effluents the commonly used solvents such as perchloroethylene may easily reach a level of several parts per million.

Wildlife

The undesirable effects of chlorinated pesticides and PCB are closely associated with their tendency to accumulate in the fatty tissues of animals, where they are found at concentrations rising to hundred of parts per million. By contrast, the concentrations reported for the C$_1$–C$_2$ chlorocarbons are of the order of a few parts per billion which is three to five orders of magnitude lower [43]. The significant compounds (Table 7) are chloroform and carbon tetrachloride, trichloroethylene, perchloroethylene and methyl chloroform. Methyl chloride, methylene chloride and ethylene dichloride were not detected but the analytical method used did not resolve lower MW compounds; while traces of hexachlorobutadiene were found in some marine organisms from an area known to receive industrial effluents. The compounds found in the animals were also found in a few samples of marine algae; but at the time of sampling, the analytical method for methyl chloride was not adequate to show its presence.

Human Food and Human Tissue

Some data on the distribution of low MW chlorocarbons in human food and human tissue have been reported [36, 40]; they also are shown in Table 7.

As in the wildlife samples, there is widespread distribution at the part per billion level, with a tendency to higher concentration in fatty tissues; the unexpected observation is the high concentration of chloroform, compared with the other compounds whose known production levels are much higher.

Distribution and Degradation

Entry to the Environment and Initial Distribution

Most of the known inputs occur either by direct volatilization to the atmosphere from production and use sites (including land disposal of solid residues), or by discharge in effluents to surface waters. Depending on their relative volatility, individual compounds will either preferentially evaporate to atmosphere, or be rained out back into surface waters. It is possible to calculate a theoretical partition coefficient

Table 7. Distribution of halocarbons in biological tissues. (Parts in 10^9 w/w wet tissue)

	Chloroform	Carbon tetra-chloride	Tri-chloro-ethylene	Per-chloro-ethylene	Tri-chloro-ethane
Plankton	0.02– 5	0.04– 0.5	0.05– 0.9	0.05– 2.3	0.03–10.7
Marine algae	(No analysis)	10 –15	16 –22	13 –20	10 –25
Molluscs	0.05–150	0.1 – 2.0	0.05–12	0.05– 6.4	0.05–10
Crustacea	0.05–180	0.2 – 3.0	2.6 –16	2 –15	0.7 –34
Fish flesh	5 –110	2 –10	0.8 –11	0.3 –11	0.7 – 5
Fish liver	6 – 18	0.3 –36	2 –56	1 –41	1 –15
Water birds eggs	0.7 – 29	1 –30	2.4 –33	1.4 –39	3 –30
Water birds liver	1.3 – 17.3	1 – 3	2.1 – 6	1.5 – 3.1	1 – 4
Seal liver	0.01– 12	0.2 – 4	3 – 6.2	0.05– 3.2	0.2 – 4
Seal blubber	7.6 – 22	4 –15	2.5 – 7.2	0.6 –19	8 –24
Humans fat	5 – 68	1 –20	1.4 –32	0.5 – 4.3	1 –25
Humans liver	1 – 10	1 – 4	2 – 5.8	0.4 –29	1 – 5
Dairy products	1 – 33	0.1 –10	0.3 –10	0.05–13	0.1 –10
Meat	1 – 4	7 – 9	12 –22	0.9 – 5	3 – 6
Vegetable oils	0.05– 10	0.05–18	0.05– 9	0.01– 7	0.5 –10
Bread [a]	2	5	7	1	2
Fruit and vegetables	0.05– 18	3 –15	0.05– 0.5	0.05– 1.2	1 – 4
Cereals [a]		0.1 –20			
Fluor [a]		0.2 – 1			

[a] Probably from use of fumigants

water/air (see Table 2) but practical experience suggests that evaporation rates are much higher than would be expected; this is possibly explained by activity coefficients leading to non-ideal behaviour, or co-distillation [53, 55].

There is some evidence from field observations that even the lower MW chlorocarbons show some tendency to adsorb on solids [43, 55] but this has not been quantified in laboratory studies.

It may be concluded that the atmosphere is the major pathway to be considered [16].

Breakdown in the Atmosphere

The concern over the effects of chlorofluoro-carbons on the ozone layer has led to a rapid increase in understanding of atmospheric breakdown-mechanisms, particularly those involving chlorine atoms. Although UV light plays a crucial part in the balance of the ozone layer, direct photodegradation is not an important mechanism in the disappearance of chlorocarbons in the atmosphere [39, 40].

Much more significant is reaction with hydroxyl radicals, which are themselves formed photolytically, but many other molecules, and free radicals are involved. The decomposition of Carbon Tetrachloride (and F 11 and F 12), which are the only halogenated compounds which reach the stratosphere in any quantity, is initiated there by direct photolytic production of a chlorine atom; some of the other

Table 8. Some breakdown rates of chlorocarbons

	Atmospheric		Aqueous half lives in weeks		Possible atmospheric reaction products
	OH ion rate $K_{OH} \times 10^{12}$ cm^3 molecule^{-1} see^{-1} weeks	Half life weeks	Hydrolysis	Solvolysis	
Methyl chloride	0.14	12	230	200	Phosgene, chlorine, formyl chloride
Methylene dichloride	0.1	15	23×10^7	22×10^4	–
Chloroform	0.1	15	10×10^4	14×10^6	Phosgene, chlorine
Carbon tetrachloride	<0.001	>1,000	$>10^8$	$>10^8$	Phosgene, chlorine
Vinyl chloride	2	1	–	–	–
Vinylidene chloride	4	0.5	–	–	Phosgene
Trichloroethylene	2.2	1	–	–	Phosgene, formyl chloride dichloro-acetic acid
Perchloroethylene	0.17	10	–	–	Phosgene, $Cl_2C(OH)COCl$, chlorine, tri-chloroacetic acid
Ethylene dichloride	0.22	8	$>10^3$ [a]	$>10^3$ [a]	Chloroacetaldehyde, chloroacetyl chloride, formyl chloride
Methyl chloroform	0.012	140	–	25 [a]	Phosgene, chlorine
Allyl chloride	28	0.05	–	–	Chloracetic acid, chloracetaldehyde

[a] By dehydrochlorination [37]

reactions which occur are illustrated schematically below

$$HCl \underset{OH^0}{\overset{CH_4}{\rightleftharpoons}} Cl^\bullet \overset{O_3}{\underset{O^\bullet}{}} ClO \underset{h\nu}{\overset{NO_2M}{\rightleftharpoons}} ClONO_2 .$$

The major mechanism for tropospheric destruction is attack by OH$^\bullet$, leading to abstraction of H$^\bullet$ from molecules which are saturated and incompletely halogenated

e.g. $$CH_3Cl + OH^\bullet \rightarrow CH_2Cl^\bullet + H_2O$$

or to elimination of a Cl atom if the molecule is unsaturated

e.g. $$Cl_2C = CCl_2 + OH^\bullet \rightarrow Cl_2HC - COCl + Cl .$$

Some values for OH ion reaction rates [16] are shown in Table 8 with estimates of atmospheric half lives based largely on those reaction rates. This photoinduced tropospheric hydroxyl ion attack is believed to be the dominant route for the degradation of most of the original compounds, except for carbon tetrachloride (which enters the stratosphere by turbulent diffusion).

The ultimate products of atmospheric degradation include HCl, CO and CO_2; free chlorine atoms will also be formed in addition to a number of chlorinated derivates of formaldehyde, acetic acid and acetaldehyde (see Table 8).

Many of these latter derivatives will, due to their water solubility, be rained out or adsorbed onto aerosol particles, where they will be exposed to the degradative processes existing in the aquatic environment.

Abiotic Breakdown in Water

Compounds in solution in water may be susceptible to hydrolysis or solvolysis, especially under conditions such as those in sea water when pH values are significantly higher than 7; in the latter case particularly, the reaction may more strictly be described as dehydrochlorination.

The rates of degradation (some of which are shown in Table 8) in water, for almost all the primary C_1–C_2 compounds of interest, are so low as to make this route of degradation unimportant [18, 37, 38]. There is some evidence that methyl chloroform and also tetra-, penta-, and hexachloroethane do undergo some dehydrochlorination, especially under alkaline conditions.

Many of the more polar intermediate products (chloroacetates, etc.) will themselves undergo some further hydrolysis, but this type of reaction is of little significance compared with their susceptibility to biodegradation.

Biodegradation in Water

It has long been believed that the carbon-halogen bond in chlorocarbons is not susceptible to biological transformations, especially by micro-organisms and lower plants; there is now growing evidence that this is not entirely true. A wide range of naturally occurring halogenated compounds is now known [Siuda 48, Volume IA of this Handbook), in addition to the already discussed methyl chloride; the soil fumigants methyl bromide, dibromobutane, dibromochloropropane and ethylene dibromide are degraded by soil micro-organisms [11], and methylene chloride has been shown to act as sole carbon source to bacteria [9].

Ultimate biodegradation of the halogenated aliphatic acids, aldehydes, etc. such as mono-, di- and tri-chloroacetic which result from the metabolism of chlorocarbons by mammals, and from some atmospheric chemical reactions, is well established [43].

Metabolism by Vertebrates

The volatile chlorocarbons are easily inhaled by mammals in the laboratory, and humans are frequently exposed to them at work and at home; after exposure, most of the material is normally rapidly lost in expired air but some metabolities and their conjugates are found in the urine. Chlorinated acetic acids, chloroethanol, chloroacetaldehyde, chloral and its hydrate, chloroglycol and chloroglycollic acid have all been identified, many of them after conjugation with glutathione; in the case of the chlorinated ethylenes which have been extensively studied, the first stage

Table 9. Effects on living organisms

	Man TLV ppm[a]	Rat LD$_{50}$ mg/kg	Fish LC$_{50}$ mg/l	Barnacle nauplius LC$_{50}$ mg/l	Alga EC$_{50}$ mg/l
Methyl chloride	100	–	700[b]		
Methylene dichloride	200	(> 2,000)[c]			
Chloroform	10	2,200	28		
Carbon tetrachloride	10	2,900	50		
Vinyl chloride	–[d]	–	ca. 3,000[b]		
Vinylidene dichloride	10	–	250		
Trichloroethylene	100	> 3,000	16	20	8
Tetrachloroethylene	100	(> 4,000)[e]	5	3.5	10.5
Ethylene dichloride	10	680	115	186	340
Methyl chloroform	350	>10,000	33	7.5	5

[a] Data from TLVs 1980 (ACGIH)
[b] Approximate result by rapid bioassay
[c] Rabbit LD$_{50}$
[d] Levels for vinyl chloride are being discussed separately
[e] Mouse LD$_{50}$

of the metabolic process is believed to be an epoxide across the double bond [13, 25, 27, 28, 32, 36, 41]. No tendency to accumulate in fatty tissues has been reported.

In the case of fish, exposed to solutions in water, some accumulation occurs, especially in the liver, bioconcentration factors reported range from 2 for ethylene dichloride in whole fish to 400 for perchloroethylene in liver, followed by very rapid loss after exposure ceased [6, 43]. No evidence for metabolism has been found but the possibility should certainly not be excluded.

Effects on Living Organisms

Mammals

All the compounds act, by inhalation, as reversible anaesthetics or intoxicants; in high acute doses, or over a longer period, they cause liver damage. As a guide to effective dose levels, Table 9 presents data on TLV's and acute injection LD$_{50}$'s.

A great deal of attention has recently been focussed on the carcinogenic potential of the class; except for vinyl chloride, the evidence is still not clear [19, 20, 31, 45].

Aquatic Organisms

Toxicity to aquatic organisms is also low [43], data on fish, barnacle nauplii and unicellular algae are presented in Table 9.

Terrestrial Species

Our knowledge of effects on terrestrial species is largely derived from the use of methyl chloride, methyl bromide, chloroform, carbon tetrachloride, ethylene dichloride, ethylene dibromide, dibromo-chloropropane and dibromobutane as fumigants, for the control of pests of stored grain, and of soil nematodes [11, 40]. A typical fumigant dose is of the order of 50 mg/l, which is similar to the measured action levels against aquatic organisms shown in Table 9.

Microorganisms

Chloroform, methyl chloroform, and to some extent carbon tetrachloride, trichloroethylene and perchloroethylene inhibit the anaerobic sludge digestion stage of biological sewage treatment, at levels as low as 5 mg/kg on dry sludge solids [8]; but there are no reports of any inhibition or aerobic microbial activity at realistic levels of occurrence.

Possibilities for Control of Discharges

There are two major possibilities for reducing man made emissions of halocarbons to the environment. The first is to reduce waste discharges after production and use, the second is to restrict the levels of THM in potable water supplies.

A limit of 100 µg/l of THM in potable water is being proposed in the USA; this could be achived by replacing chlorine as a disinfectant for water by other agents, by a radical change in chlorination practices, or by adsorption of THM after chlorination by active carbon or synthetic resins. If the latter procedure is adopted, the spent carbon or resin itself then becomes a waste to be treated.

The commonest method for the purification of liquid effluents including sewage is biological treatment – during the aerobic stage, the low MW halocarbons are probably lost to atmosphere, but not biodegraded. Similarly, chlorinated solvents present in solid waste which is disposed of to landfill will evaporate, over a period depending on the nature of the waste and the type of site operation.

Physical methods of treatment of effluents, liquid or gaseous, such as adsorption, flocculation and sedimentation will concentrate halocarbons on the separated solid phase. The only reliable method of destruction is controlled incineration, at high temperature.

General Overview

Chlorinated hydrocarbons, as a broad class, are at present subject to massive scrutiny by regulatory authorities throughout the world. They are on the "Black List" of all the International Conventions regulating the discharge of wastes to the aquatic environment; many of them are specifically listed in the Toxic Pollutants list of the USA Clean Water Act, are ITC candidate compounds for Section 4 Test

Rules under TSCA in the USA, and are considered for listing for control under the Clean Air Act, and under toxic waste regulations. As a result, a number of assessments have been made, such as that of the Halomethanes [40], and more are in progress.

There are many gaps in our knowledge, but it is clear that a large quantity of chlorinated C_1 and C_2 compounds is released to the environment, where they are rapidly dispersed, largely to the atmosphere. Breakdown mechanisms are dominated by photoinduced tropospheric degradation, followed by some microbial biodegradation of the more polar by-products in water.

Many of the compounds, notably methyl chloride, chloroform, carbon tetrachloride, trichloroethylene, perchloroethylene and methyl chloroform are found widely distributed, at low concentrations, in air, water, sediments, and in the tissues of plants, animals and man himself; these concentrations are well below those that could cause any proven adverse biological reactions.

The absence of reports of the presence in the environment of vinyl chloride, vinylidene chloride, methylene chloride, ethylene dichloride, propylene dichloride, allyl chloride, epichlorhydrin and chloroprene, and any of the more polar intermediate degradation products found in laboratory studies, suggest that the natural degradation processes of these compounds proceed at higher rates than inputs.

The possibility of interference with the ozonosphere is discussed elsewhere.

The paucity of data in some areas, especially a clearer understanding of the importance of the natural production and cycling of low molecular weight halocarbons, is inhibitng further decision making.

References

1. Ahnoff, M., et al.: Water Res. *13*, 1233 (1979)
2. Anon.: Air Pollution form Chlorination Processes, EPA PB-218-048 (1972)
3. Appleby, A.: Atmospheric freons and halogenated compounds, EPA-600/3-76/108 (1976)
4. Berge, G., Ljøen, R., Palmork, K.H.: The disposal of containers with industrial wastes into the North Sea. FAO Conf., Rome (1970), Paper E-73
5. B.I.T. (Solvents Chlores): Anal. Chim. Acta. *82*, 1 (1976)
6. Barrows, M.E., et al.: Bioconcentration and elimination of selected water pollutants by Bluegill Sunfish. Dynamics, Exposure and Hazard assessment of Toxic Chemicals [Hague, R. (ed.)] (1980)
7. Bauer, U., Selenka, A.: Analysis, presence and behaviour of halogenated hydrocarbons in drinking water preparation, Spez. Ber. Kernforschungs-Anlage Jülich – Issue Jül.-Spez-45, Organohalogenverbindungen Umwelt (1979)
8. Brown, D.: Chlorinated solvents in sewage works effluent and water treatment, March 1978
9. Brunner, W., Leisinge, T.: Experentia *34*, 1671 (1978)
10. Callahan, M.A., Ehreth, D.J., Levins, P.L.: Sources of toxic pollutants found in influent to sewage treatment plants, Proc. Natl. Conf. Munic. Sludge Managem. 8 th (1979) p. 55
11. Castro, C.E.: Environm. Health Perspect. *21*, 279 (1977)
12. Croll, B.T.: The DOE Haloforms Survey. Analytical Considerations in Trihalomethanes in Water – 1980, Seminar – Water Res. Centre, U.K.
13. Duuren, B.L. van: Environm. Health Perspect. *21*, 17 (1977)
14. E.P.A.: Major Organic Products, Devel. Doc. Eff. Limitation Guidelines, EPA 440/1-73/009 (1973)
15. E.P.A.: Organic Chemicals Industry Phase II, Devel. Doc. Eff. Limitation Guidelines, EPA (1974)
16. E.P.A.: Fate of Toxic and Hazardous Materials in the Air Environment, EPA-600/53-80-084 (1980)
17. Feiler, H.D., Vernick, A.S., Storch, P.J.: Fate of Priority Pollutants in POTWs, Proc. Natl. Conf. Munic. Sludge Manage. 8 th (1979), p. 72

18. Fells, I., Moelwyn-Hughes, E.A.: J. Chem. Soc. *1959*, 398
19. Fishbein, L.: Sci. Total Environm. *7*, 111 (1979)
20. Fishbein, L.: ibid. *7*, 163 (1979)
21. Giger, W., Molnar-Kubica, E., Wakeham, S.: Volatile chlorinated hydrocarbons in ground and lake waters. Transformation and biological effects. Pergamon Ser. Environ. Sci. (1977), VI Aquatic Pollutants [Hutzinger (Ed.)]
22. Glaze, W.H., Henderson, J.E., IV, Smith, G.: Analysis of new chlorinated organic compounds in municipal waste waters after terminal chlorination, in: Identification and Analysis of Organic Pollutants in Water [Keith (Ed.)], Ann Arbor (1977), p. 247
23. Hanst, P.L., et al.: J. Air Pollution Control Assoc. *25*, 1220 (1975)
24. Helz, G.R., Hsu, R.Y.: Limnol. Oceanogr. *1978*, 858
25. Henschler, D.: Environm. Health Perspect. *21*, 61 (1977)
26. Harris, G.: Perchloroethylene and trichloroethylene, Technical and Commercial aspects, Proc. 5th Intern. Conf. Europ. Chem. Marketing Res. Assoc. 1971
27. Hathway, D.E.: Environm. Health Perspect. *21*, 55 (1977)
28. Ikeda, M.: ibid. *21*, 239 (1977)
29. Jolley, R.L., et al.: Determination of chlorination effects on organic constituents in natural and process waters, in: Identification and Analysis of Organic Pollutants in Water [Keith (Ed.)], Ann Arbor (1977)
30. Kaiser, K.L.E., Valdmanis, I.: Volatile Chloro- and chlorofluorocarbons in Lake Erie 1977/78, J. Gt. Lakes Res. *5*, 160 (1979)
31. Lee, C.C., et al.: Environm. Health Perspect. *21*, 25 (1977)
32. Leibman, K.C., Ortiz, E.: ibid. *21*, 91 (1977)
33. Löchner, F.: Münchner Beitr. Abwasser, Fisch, Flussbiol., Issue: Schadstoffe im Oberflächenwasser und im Abwasser *30*, 227 (1978)
34. Lovelock, J.E.: Nature *241*, 194 (1973)
35. Lovelock, J.E., Simmons, P.: Halocarbons in the atmosphere in Chlorofuorocarbons in the Environment, in: The Aerosol Controversy [Sugden (Ed.)], Sci-Horwood (1980)
36. McConnell, G., Ferguson, D.M., Pearson, C.R.: Endeavour *34*, 13 (1975)
37. McConnell, G.: unpublished
38. Moelwyn-Hughes, E.A.: Proc. Roy. Soc. A *220*, 386 (1953)
39. N.A.S.: Halocarbons. Effects on Stratospheric Ozone, US Nat.. Acad. Sci. (1976)
40. N.A.S.: Chloroform, Carbon Tetrachloride, and other halomethanes. An Environmental Assessment. US Nat. Acad. Sci. (1978)
41. Ogata, M., Norichika, K., Shinada, Y.: Acta Med. Okayama *33*, 415 (1979)
42. Palmer, T.Y.: Nature *263*, 44 (1976)
43. Pearson, C.R., McConnell, G.: Proc. Roy. Soc. Lond. B. *189*, 305 (1975)
44. Pearson, C.R.: unpublished
45. Rampy, L.W.: Environm. Health Perspect. *21*, 33 (1977)
46. Rasmussen, R.A. (ed.): 19 Rep. Workshop on Halocarbon Analysis and Measurement, NASA-NBS-NSF Boulder Colorado, March 1976
47. Singh, H.B., Salas, L.J., Cavanagh, L.A.: J. Air Poll. Control Assoc. *27*, 322 (1977)
48. Siuda, J.F., DeBernardis, J.F., Cavestri, R.C.: Gas chromatographic detection of naturally occurring halogenated compounds derived from marine organisms. Food-Drugs from the sea. Marine Technol. Soc. Proc. 1972 [Worthen, L.R. (ed.)]
49. Sonneborn, H.M. (ed.): Gesundheitl. Probleme der Wasserchlorung und Bewertung der dabei gebildeten halogenierten organischen Verbindungen, WABOLU-BER 3/1978
50. Stanford Research Institute: unpublished
51. Su, C., Goldberg, E.D.: Marine Pollutant Transfer (1976), p. 353
52. Fielding, M.: Formation of trihalomethanes in: Trihalomethanes in Water Seminar. Water Research Centre U.K. 1980
53. Mackay, D., Wolkoff, A.W.: Envir. Sci. Technol. *7*, 611 (1973)
54. Rook: Water Treatment & Examination *23*, 234 (1974)
55. Scherb, K.: Münch. Beiträge Abwasser, Fisch, Flussbiol. *30*, 235 (1978)

Halogenated Aromatics

C. R. Pearson

Imperial Chemical Industries Ltd., Brixham Laboratory
Overgang Brixham, Devon TQ5 8BA, England

Introduction

Apart from pesticides, chlorinated aromatic compounds have attracted, and continue to attract, more attention from scientists, legislative bodies, and pressure groups than any other class of compound. Polychlorinated biphenyls (PCB) are by far the best known members of the class, but many other chlorinated products, and some brominated materials, have considerable commercial significance. Fluorine and iodine derivates are of little importance, and will not be considered.

The more heavily chlorinated polycyclic compounds are commercially valuable because of their stability, especially at high temperatures, good hydraulic and electrical properties, and fire resistance. The brominated analogues have similar properties, but are mainly used because of their fire retardant characteristics. This stability is associated with the less desirable aspects of their environmental behaviour, and together with their physical properties can lead to accumulation in fatty tissues, low environmental mobility, and resistance to degradation, particularly biodegradation.

The less chlorinated benzenes and substituted benzenes are more reactive, and their major use is as intermediates in chemical manufacture. The related chlorinated phenols are considered in Vol. 3A of this Handbook.

This literature on PCBs is very extensive; a number of valuable reviews and conference proceedings have been published [3, 4, 6, 14, 35, 42, 75, 78, 82, 95, 96, 98, 100, 104, 115, 133]. That on the other compounds is much less full; but reviews are available on PCN [30, 63], PBB [22, 31] and chlorobenzenes [28, 32, 34, 134].

Production and Properties

All the compounds covered in this chapter are produced by direct halogenation of the parent hydrocarbon (or substituted hydrocarbon).

Estimates of world production are given in Tables 1 and 2, together with major uses. Although these are not the only factors involved, they do help to place the compounds in an order of potential priority as environmental contaminants. Available data on composition and properties are presented in Tables 3 and 4.

Polychlorinated Biphenyls (PCB)

PCBs are produced by the direct anhydrous chlorination of biphenyl, using iron filings or ferric chloride as a catalyst; when chlorination has reached the required

Table 1. Production and uses of halogenated aromatic compounds (more than one ring) [22, 24, 63, 65, 81, 95, 133, 141]

Compound	Major trade names	World production 1,000 tonne/ year	Major uses
PCB	Aroclor Phenoclor Kanechlor Clophen Fenclor Santotherm	70[a]	Capacitor dielectric, transformer coolant, hydraulic fluid, plasticiser, heat transfer fluid
PCT	Aroclor Kanechlor	5?	Adhesives and sealants
PCN	Halowax	5?	Capacitor dielectric, oil additive
PBB	Firemaster	5[b]	Fire retardant

[a] Peak annual rate for 1969–1970
[b] Peak annual rate for 1974

Table 2. Production and uses of chlorinated aromatic compound (Benzene derivatives) [105]

Compound	World production 1,000 tonne/year	Major uses
Chlorobenzene	600	Solvent, intermediate
1,2 Dichlorobenzene	80	Solvent, intermediate
1,4 Dichlorobenzene	80	Odoriser, insect repellent
1,2,3 Trichlorobenzene	1	Intermediate
1,2,4 Trichlorobenzene	30	Intermediate
1,3,5 Trichlorobenzene	1	Intermediate
1,2,4,5 Tetrachlorobenzene	5	Intermediate
Hexachlorobenzene	10	Fungicide, by-product
Pentachloronitrobenzene	4	Fungicide
1-Chloro, 4-Nitrobenzene	20	Intermediate
2-Chloro Aniline	15	Intermediate
4-Chloro Aniline	5	Intermediate
2,5-Dichloro Aniline	2	Intermediate
3,4-Dichloro Aniline	20	Intermediate
Chloro Toluene	10?	Intermediate

Table 3. Composition and physical properties (compounds with more than one benzene ring) [12, 22, 24, 63, 65, 81, 95, 99, 141]

Name	Appearance	M.W. (Avg)	% Cl	Sp. Gr.	Distillation range °C	Water solubility μg/l	Vapour pressure mm Hg 20 °C	Carbon adsorption[c] mg/g
PCB								
Aroclor 1016	Mobile Oil	258	41	1.400	325–350	220– 900	2×10^{-4}	242
1221	Mobile Oil	192	21	1.185	275–320	3,500–5,000	10^{-2}	630
1232	Mobile Oil	221	32	1.275	290–325		$5 \times 10^{-3} - 10^{-3}$	
1242	Mobile Oil	261	42	1.385	325–366	200– 700	$9 \times 10^{-4} - 4.1 \times 10^{-4}$	
1248	Mobile Oil	288	48	1.410	340–375	54– 100	$8.3 \times 10^{-4} - 5 \times 10^{-4}$	
1254	Viscous liquid	327	54	1.500	365–390	12– 70	$1.8 \times 10^{-4} - 7.7 \times 10^{-5}$	
1260	Sticky resin	372	60	1.560	385–420	3– 25	$0.9 \times 10^{-4} - 4.1 \times 10^{-5}$	
1262	Sticky resin	389	62	1.580	390–425			
1268	Whitish powder	423	68	1.810	435–450			
PCT								
Aroclor 5460	Brittle resin	575	60	1.67	280–335 (at 5 mm Hg)			
Chloronaphthalenes								
Halowax 1031	Mobile Oil	151	22	1.2	250–276		10^{-1}	280 (2-chloro-)
1000	Mobile Oil	159	26	1.22	250–300		5×10^{-2}	
1001	Flakes	207	50	1.58	308		ca. 5×10^{-4} [a]	
1099	Flakes	209	52	1.59	318			
1013	Flakes	225	56	1.67	328		ca. 10^{-4}	
1014	Flakes	246	62	2.78	344			
PBB								
Firemaster BP-6	White powder	556	75[b]	2.57		10– 30	ca. 10^{-6}	

[a] Extrapolated from [63]
[b] % Bromine
[c] [23]

Table 4. Composition and physical properties (Benzene derivatives) [28, 32, 33, 79, 105, 131, 134]

Name	Formula	Molecular Wt.	% Chlorine	No Cl Atoms	MP °C	BP °C	Vapour pressure mm Hg @ 20 °C	Solubility in water mg/l @ 20 °C	Partition Coefficient log $P_{o/w}$	Carbon adsorption mg/g[a]
Chlorobenzene	C_6H_5Cl	113	32	1	− 45	132	8.8	500	2.84	91
1,2 Dichlorobenzene	$C_6H_4Cl_2$	147	48	2	− 17.5	179	1	100	3.55	129
1,3 Dichlorobenzene	$C_6H_4Cl_2$	147	48	2	− 25	172	3	123	3.44	118
1,4 Dichlorobenzene	$C_6H_4Cl_2$	147	48	2	53	173	0.6	79	3.38	121
1,2,3 Trichlorobenzene	$C_6H_3Cl_3$	182	58	3	53	219	< 1		4.11	
1,2,4 Trichlorobenzene	$C_6H_3Cl_3$	182	58	3	117	213	< 1		3.93, 4.23	
1,3,5 Trichlorobenzene	$C_6H_3Cl_3$	182	58	3	63	208	< 1	30	4.15	151
1,2,3,4 Tetrachlorobenzene	$C_6H_2Cl_4$	216	66	4	47.5	254			4.46	
1,2,3,5 Tetrachlorobenzene	$C_6H_2Cl_4$	216	66	4	51	246			4.93, 4.5	
1,2,4,5 Tetrachlorobenzene	$C_6H_2Cl_4$	216	66	4	138	246	< 1		4.52	
Pentachlorobenzene	C_6HCl_5	261	72	5					5.6, 4.88	
Hexachlorobenzene	C_6Cl_6	285	75	6	228	325	10^{-5}	.007	6.0, 5.0	450
Pentachloronitrobenzene	$C_6Cl_5NO_2$	296	63	5	146					
1-Chloro 4-Nitrobenzene	$C_6H_4ClNO_2$	158	23	1					2.4	130
2-Chloro aniline	$C_6H_4ClNH_2$	128	28	1	− 14/− 3.5	209			1.9	
3-Chloro aniline	$C_6H_4ClNH_2$	128	28	1	− 10.4	230	< 0.1		1.9	
4-Chloro aniline	$C_6H_4ClNH_2$	128	28	1	70–72	230–231	.015	5,300	1.8	
2,3-Dichloro aniline	$C_6H_3Cl_2NH_2$	162	44	2					2.8	
2,5-Dichloro aniline	$C_6H_3Cl_2NH_2$	162	44	2	50	251				
2,6-Dichloro aniline	$C_6H_3Cl_2NH_2$	162	44	2						
3,4-Dichloro aniline	$C_6H_3Cl_2NH_2$	162	44	2	71.5	272			2.7	
3,5-Dichloro aniline	$C_6H_3Cl_2NH_2$	162	44	2						
2-Chloro toluene	$C_6H_4CH_3Cl$	127	28	1	− 36/− 34	159	2.7		3.42	

[a] [23]

stage, the crude product is purified by alkali wash, followed sometimes by distillation [6, 95].

The commercial products are all mixtures of isomers, of which 209 are theoretically possible; although most of these have been synthesized, only a proportion, consisting of those whose chlorine distribution is reasonably symmetrical, are found to any significant extent [65]. Commercial PCB is sold as different grades, based on chlorine content; they are often blended to match specification, both in terms of chlorine content and quality. The principal manufacturer, in the USA, Monsanto, (Trade name Aroclor) uses a number code system to indicate chlorine content (e.g. 1242 in 42% chlorine). The salient characteristics of the Aroclor grades, used as an example, are shown in Table 3; those of other manufacturers' products will not be significantly different.

The distribution of numbers of isomers, and their occurrence in commercial grades of Aroclor, are shown in Tables 5 and 6. The most important grade in terms

Table 5. Chlorinated biphenyls – composition and properties of isomer groups [65, 95]

Number of chlorine atoms	Number of isomers	M.W.	% Cl	Solubility in water µg/l
0	1	154	0	
1	3	189	19	1,100–5,900
2	12	223	32	80–1,400
3	24	257	41	78– 85
4	42	291	49	34–ॱ180
5	46	326	54	22– 31
6	42	361	59	88
7	24	395	63	
8	12	430	66	7
9	3	464	69	
10	1	499	71	15

Table 6. Isomer composition of PCB grades [65, 95]

Number of chlorine atoms	Aroclor Grade					
	1016	1221	1232	1242	1254	1260
0		11	6			
1	1	51	26	1		
2	20	32	29	16	0.5	
3	57	4	24	49	1	
4	21	2	15	25	21	
5	1	0.5	0.5	8	48	12
6				1	23	38
7					6	41
8						8
9						1
10						

of quantity produced is 1242, which is predominantly tri-chloro-, with a large proportion of tetrachloro-; 1061 is a later version of 1242, with reduced levels of tetra- and pentachloro-. The other major products are 1221 (largely monochloro-) and 1254 (mainly pentachloro-). Few commercial products contain any significant amounts of octa-, nona- or decachloro- material.

Some commercial products consist of mixtures with other compounds such as chlorobenzenes.

Polychlorinated Terphenyls (PCT)

PCT are prepared by direct anhydrous chlorination of terphenyl, as for PCB. Commercial grades are similarly mixtures of isomers; the possible number is very large, as the terphenyl structure itself has 3 isomeric forms (o-. m. and p-). (Fig. 1). The grade normally used is Aroclor 5460 (or its Kanechlor equivalent), which contains ca. 60% chlorine; it is often sold in admixture with PCB grades. Basic properties are summarised in Table 3.

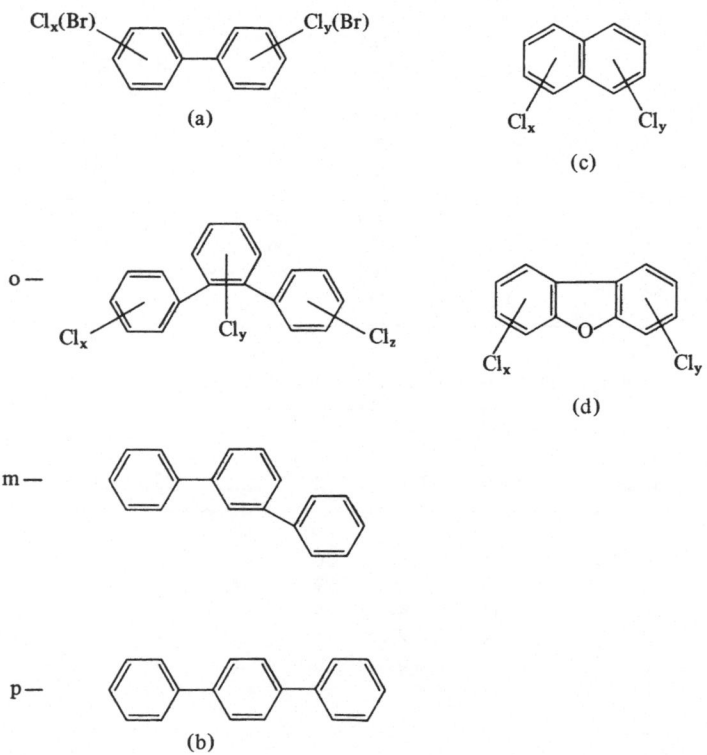

Fig. 1. Structures of polycyclic compounds. (a) Polychlorinated (brominated) biphenyl; (b) Polychlorinated (o-, m-, p-) terphenyl; (c) Polychlorinated naphthalene; (d) Polychlorinated dibenzofuran

Table 7. Isomer composition of PCN grades [63, 81]

Number of chlorine atoms	Halowax grade						
	1031	1000	*1001*	*1009*	1013	1014	1061
1	95	60					
2	4	40	10				
3			40		10		
4			40		50	20	
5			10		40	40	
6						40	
7							10
8							90

Chlorinated Naphthalenes (PCN)

The result of direct anhydrous chlorination of naphthalene, using ferric or antimony chloride as catalyst, is a series of isomers, ranging from mono- to octa-chloro-(Fig. 1); 79 iosomers are theoretically possible, most of which have been isolated. The basic properties of the main commercial grades are shown in Table 3; the isomeric composition of these is shown in Table 7.

Impurities in PCB, PCT, PCN

All the commercial grades are by definition impure, as they are a mixture of substances. Most manufacturers offer both "technical" and "purified grades;" the only significant difference appears to be colour.

Anomalies in toxicological test results and in particular the Yusho poisoning outbreak in Japan, led to a more intensive search for impurities. As many of the parent hydrocarbon feedstocks are derived from oil or coal tar sources, they tend to be contaminated by other aromatics, especially each other; so PCN may be found in PCB, and vice versa, but at levels well below 0.1%.

The best documented contaminants are the chlorinated benzofurans [59, 95, 98], containing anywhere from 1 to 8 chlorine atoms [Fig. 1(d)]; these have been found in commercial PCB samples, typically at the 1–10 mg/kg level. There is evidence that the concentration rises during service at high temperatures.

The Yusho rice oil which contained high levels of PCDF has recently been reported as containing significant quantities of chlorinated terphenyl, and also of chlorinated tetraphenyl [94].

Polybrominated Biphenyls

PBBs are normally prepared by direct bromination of biphenyl, using an aluminium chloride catalyst; other manufacturing routes, using Friedal Craft processes, are possible [97]. The principal grade produced was basically a hexabromo- (Firemaster BP-6) whose approximate composition was given as tetra- 2%, penta- 10%, hexa- 63%, hepta- 14%, other 11%. Octo-, deca-, and some nona- may also be produced in small quantities.

Little information on properties is available; it is shown in Table 3. Some contamination by brominated naphthalenes has been reported.

Chlorinated Benzenes

These are prepared by successive chlorination of benzene, or mono- or dichlorobenzenes, using ferric or aluminium chloride catalysts; all the isomers may be separated, but are often sold as commercial grades, eg mixtures of di- and tri-isomers. The important products are mono-, o- and p-, di-, and grades of tri- and tetra-; meta-dichlorobenzene is not formed in any significant proportion.

Hexachlorobenzene is found in the "heavies" fractions of many chlorination processes, in addition to chlorobenzenes production, from which it may be isolated if necessary.

The other substituted products all include a chlorination stage as part of their manufacturing process.

The insecticide BHC (benzene hexachloride, Lindane) is also produced by the chlorination of benzene, but under conditions which favour chlorine addition not substitution, using U.V. light as a catalyst. It is not an aromatic compound at all, but a saturated hexachlorocyclohexane. The fact that it has properties which are similar to those of hexachlorobenzene, and that it is detected by the same analytical methods for the chlorinated pesticide group, often causes confusion.

Uses and Losses to the Environment

Polychlorinated Biphenyls

PCBs were first produced in 1929, in USA; annual production and use increased steadily until 1970, and became effectively worldwide. Uses can be classified into 3 types:
(a) Closed electrical – transformers, capacitors, etc. It is assumed that leakages are minimal, and that the lifetimes of the components are long (e.g. > 20 years)
(b) Nominally closed – hydraulic and heat transfer fluids. Leakages must be expected, and lifetime between changes is much shorter (ca. 1 y)
(c) Dispersive – plasticisers, paint and printing ink components, adhesives, and additives for cutting oils, textile auxiliaries and pesticides. Material is rejected almost immediately after use.

Following the growing number of observations of widespread environmental occurrence Monsanto in 1971 ceased the sale of PCB for all applications except electrical and some hydraulic; this voluntary ban was followed by other producers, and by 1977 worldwide manufacture for any purposes had virtually ceased.

Until then, entry of PCB to the environment had been by the following routes:
(a) Completely dispersive losses – all material used in paints, inks, adhesives, cutting oils, etc must be assumed to have been lost, via widely scattered sources, either direct to waterways, or indirectly from land disposal.

(b) Material used for hydraulic and heat transfer fluids, and small capacitors and transformers, will largely have been discharged to the environment as under (a) but from a smaller number of point sources.

(c) Losses from manufacture of PCB, and from transformer filling and emptying operations, will have entered waterways in the immediate vicinity of these plants, with some being lost to the atmosphere.

(d) The bulk of the material used for transformers and capacitors will either still be in service, or have been disposed of in large batches, often still in the equipment, on landfill sites.

Total cumulative world production of PCB has been estimated at about 750,000 tonnes; peak production rate was about 70,000 tpa in 1970. Of these totals, approximately 60% went into closed electrical uses, 15% to nominally closed, and 25% to dispersive uses. It may be deduced therefore that some 40%, that is 300,000 tonnes, has entered the environment since 1929 in widely disseminated form, and that 60%, 450,000 tonnes, is either still in service or in landfills. Mass balances calculated for the USA are in broad agreement with these estimates [95]. They confirm that the current rate of input, via domestic sewage and industrial waste, is now down to a few hundred tonnes per year worldwide.

Two adventitious sources which have been suggested as being responsible for new input are side reactions from chlorination. The first relates to a range of synthetic chlorination processes in which hydrocarbon feedstock is present, which could generate some PCB from biphenyl present as impurity, or even from benzene; the case normally quoted is that of dyestuffs, particularly phthalocyanines. Recent studies by dyestuffs manufacturers have shown that PCB levels in dyes are now well below 50 mg/kg [2]. The second is production during the chlorination of sewage, or industrial effluents [45]; the occurrence of PCB in a range of effluents that have not been chlorinated, and of absence in those that have [51] suggests that this is not a significant source.

Polychlorinated Terphenyls

Little information is available on the production and disposal of PCT. Manufacture in Japan commenced in 1954, and was apparently still current in 1977 [24]. Major uses are in adhesives, sealants, etc, often in combination with PCB, so it may be assumed that annual production is equivalent to annual input to the environment; a maximum figure of 5000 tonnes per annum is suggested which is similar to the commonly quoted 1–20 ratio with PCB [24]. There is some doubt as to whether PCT are still marketed.

Chlorinated Naphthalenes

Production started in the late 1930s, reached a maximum of perhaps 5,000 tonnes/annum and is apparently falling [63, 81]. Uses are mainly as small capacitor dielectric, and oil additives, all of which are effectively dispersive, so annual production may be taken as annual input to the environment.

Polybrominated Biphenyls

PBBs were only produced in any significant quantity between the years 1970 and 1974, when manufacture in the USA ceased following the Michigan cattle food contamination incident [47, 66]. Until then, some 10,000 tonnes had probably been produced worldwide, for use as fire retardants in plastics and foams; this must be regarded as having by now been dispersed in the environment, partly in areas of high concentration around producer and processing plant, and in Michigan, with the rest widely distributed [97].

Chlorinated Benzenes

Production and major uses of the commercially important chlorobenzenes are shown in Table 2; most compounds are used mainly as intermediates in the synthesis of fine chemicals, and will only enter the environment as losses and wastes from the production sites. As losses are normally 1%–2% of raw material charges, their total quantities will not exceed a few hundred tonnes/annum. The exceptions are the dispersive uses of mono- and the 2 dichloro-benzenes, which have been estimated at 300,000 tonnes/annum for mono-, 40,000 tpa for o-di-, and 80,000 tpa for p-di-; of PCNB as a fungicide 4,000 tpa; and of waste disposal, and the few remaining fungicide uses of HCB, 10,000 tpa [134].

In spite of their comparatively high boiling points and water solubilities, the lower chlorinated benzenes are very volatile, so a large proportion of the losses will enter the atmosphere, either directly or indirectly.

Methods of Analysis

The most widely used method of detection and quantification of halocarbons is gas chromatography, normally with an electron capture detector. In some cases, air or water samples can be injected directly into the instrument; where initial levels are low, a concentration step consisting either of adsorption onto active carbon, or solvent extraction, may be necessary.

Much of the chloro-organic load in air samples is carried on the particulate constituents, dusts or aerosols; care is needed in separating these, if necessary, on filters of known porosity [42, 87, 91]. The same is true of particulate matter in natural water samples.

As in any GC process, identification relies on comparing retention time on the column with that of a standard; use of the electron capture detector, which is particularly sensitive to halogens, gives a greater measure of confidence to this process. But because of this high sensitivity, there is a wide range of response to different chlorinated compounds depending on the number of chlorine atoms in the molecule; there is a marked fall in sensitivity below about 80% chlorine content, and in some situations it may be better to use the more rugged flame ionisation detector.

Where analysis of sediments, foods or biological tissues is required, it is necessary to carry out a solvent extraction, followed by clean-up of the extract to remove non-halogenated fatty material. Great care in operation, and the use of highly purified solvents, are needed [21, 37, 65, 91, 96, 135].

The number of interfering compounds present in many samples, particularly of sludges, sediments and biological tissue is very large, and may make identification by GC alone extremely difficult, especially in the case of unexpected compounds. Greater certainty in identification, particularly in the case of unknown sources, may be obtained by the use of a mass spectrometer, used in series with gas chromatography [73, 94, 121]. It has a further advantage in being equally sensitive to all compounds, regardless of the chlorine content. While GC/EC ranges in sensitivity from about 5 parts in 10^5 for chlorobenzene to 1 part in 10^{12} for hexachlorobenzene, GC/MS will give a routine sensitivity of 1 part in 10^9 for all compounds.

Full schemes for the identification of large numbers of trace contaminants in waters and effluents, using concentration, separation, and GC/MS analysis have recently been evolved [29, 33].

A particular problem for products such as PCB, PCT, PCN is that in GC traces, there is no single peak representing a product, but one for each significant isomer; although such peak patterns tend to be quickly recognisable, quantification is difficult. The alternatives are either to report each separate isomer, which is a laborious process, or to compare with selected standards, using arbitrary methods to integrate all the isomer peaks; the PCB grades normally used are Aroclor 1242 and 1254.

Another approach is to perchlorinate the sample under controlled conditions, and then measure the resulting decachlorinated-biphenyl, or perchloro-terphenyl. Some of the higher molecular weight, higher boiling halogenated impurities and by-products are not volatile under GC conditions; these may be determined by high performance liquid chromatography [11, 12, 74].

There is a growing interest in the quantification of total organic halogens in environmental samples. Traditional estimates have always been made by adding up all the individual identified components; doubt is being cast on this procedure by the growth in the number of previously unsuspected compounds which are now being identified, especially those of natural origin. The best known method involves pyrolysis followed by micro-coulometry, but neutron activation techniques may also be employed [1, 51].

Much of the total organochlorine content, even in remote areas such as the polar regions, still consists of chlorinated insecticides; in spite of popular belief, DDT is still used in large quantities, particularly in the southern hemisphere, and the production, use, and frequency of detection of toxaphene are still growing [7].

Occurrence in the Environment

Polychlorinated Biphenyls

The literature on the occurrence of PCB in environmental samples is vast; a summary of typical ranges reported in air, water and sediment is presented in Table 8, and in biological tissues in Table 9. The data are mainly derived from review articles [78, 95, 132], but also include some of the classical early observations [72, 108, 113, 114, 140] and some more recent reports [26, 27, 46, 55, 61, 71, 76, 88, 109, 110, 116, 130, 136].

It is unfortunate that the objective behind most of the studies appears to be an enthusiasm for analysis, rather than a coherent planned approach to an understanding of the behaviour of chlorinated aromatic compounds in the environment; it is therefore surprisingly difficult to determine patterns or trends, either spatial or temporal, on other than a very qualitative basis.

The levels listed in Tables 8 and 9 are rounded off to illustrate the order of magnitude of occurrence in different phases of the environment; most reported data will lie inside the ranges quoted, but some figures are much higher. From the individual results, some generalisations can be made. PCB concentrations in air, surface fresh water, and river, estuary and coastal sediments are much higher near urban areas than in rural and oceanic; but PCB is clearly detectable in remote mountainous and polar regions [7]. High concentrations are also found in indoor air, due in particular to electrical uses [90].

In biological tissues, the highest levels occur in organs rich in fat, and particularly in storage fat. The concentrations found vary on a local and seasonal basis, and with the life-stage of the organism [106]; there is also considerable variation, of up to 2 orders of magnitude, between individuals of the same species collected at the same time and place. The concentrations of PCB in the tissues, particularly eggs and fat, of predatory birds and marine mammals are higher, by one or two orders of magnitude, than those of fish and invertebrates. The number of observations of levels in terrestrial animals, plants and soil is so small that no significant conclusions can be drawn. There is widespread distribution in human tissue; transfer to babies is initially due, partly to placental blood flow and partly to occurrence in maternal milk. Levels in occupationally exposed workers are higher than the population at large [19]. There is some evidence that levels in air and some waterways fell between 1970 and 1975; but there is no discernible downward trend since then, in levels observed in any environmental compartment.

Attempts to draw up a mass balance of PCB present in the environment suggest that between 25% and 75% of the total quantity produced, excluding that remaining in service, is still in existence; the distribution pattern of that present in the USA and Atlantic Ocean is as follows [95]

Atmosphere	18	tonnes
Fresh waters	12– 35	tonnes
Freshwater sediment	1,400– 7,100	tonnes
Freshwater biota	30	tonnes
Marine waters	6,000–66,000	tonnes
Marine sediment	660– 2,700	tonnes
Marine biota	30	tonnes
Soil	140– 2,800	tonnes
Sewage sludge	4,800	tonnes

Table 8. Ranges of concentrations reported in the abiotic environment

	Air ng/m³ (10⁻¹²)	Rain/Snow ng/l (10⁻¹²)	Sea water ng/l (10⁻¹²)	Surface Fresh water ng/l (10⁻¹²)	Sediments and soil μg/kg (10⁻⁹)	Sewage and Effluents μg/l (10⁻⁹)	Sludge mg/kg (10⁻⁶)
PCB	0.1–20	0.1–200	0.25–100	0.1–3,000	1–1,000	0.1–50	1–100
PCT	N.A.	N.A.	N.A.	70	5	7.5	5
PCN	25	N.A.	2–20	200	100	N.A.	N.A.
Chlorobenzene	N.A.	N.A.	N.A.	5–5,000	N.A.	0–10	N.A.
o-dichlorobenzene	N.A.	N.A.	N.A.	10–1,000	N.A.	0–100	0–10
p-dichlorobenzene	15,000	N.A.	N.A.	10–1,000	N.A.	0–100	0–10
Trichlorobenzenes	N.A.	N.A.	N.A.	10–200	N.A.	0–200	N.A.
Tetrachlorobenzenes	N.A.	N.A.	N.A.	10–100	N.A.	0–200	N.A.
Pentachlorobenzene	N.A.	N.A.	N.A.	1–25	N.A.	N.A.	N.A.
Hexachlorobenzene	Trace	N.A.	0.1–10	0.1–50	N.A.	0–100	N.A.

N.A. = No data reported; it does not imply that none is present

Table 9. Ranges of concentrations reported in biological tissues (in mg/kg wet weight)

	Plankton	Aquatic invertebrates	Fish	Birds (Adipose tissues)	Birds Eggs	Marine mammals and amphibia (blubbers)	Man (Adipose tissue)
PCB	0.01–2.0	0.01–10	0.01–25	0.1–1,000	0.1–500	0.1–1,000	0.1–10
PCT	N.A.	0.12	0.001–0.4	0.01–17	0.1–1.8	0.5–1	0.1–10
PCN	N.A.	N.A.	0.004–0.12	0.7–70	N.A.	N.A.	0.003–0.016
Chlorobenzene	N.A.	N.A.	N.A.	N.A.	N.A.	N.A.	N.A.
o-dichlorobenzene	N.A.	Tr	0.1–1.0	N.A.	N.A.	N.A.	N.A.
p-dichlorobenzene	N.A.	Tr	0.1–0.5	3.0	N.A.	N.A.	10.0
Trichlorobenzenes	N.A.	0.1–1.0	0.001–0.5	N.A.	N.A.	N.A.	N.A.
Tetrachlorobenzenes	N.A.	0.1	0.001–0.05	N.A.	N.A.	N.A.	N.A.
Pentachlorobenzene	N.A.	0.1	0.005–0.1	N.A.	N.A.	N.A.	N.A.
Hexachlorobenzene	0.0001–0.0003	0.001	0.02–20	0.01–12	0.01–3.2	0.01–1	0.2–1.5

N.A. = No data reported; it does not imply that none is present.

Although the upper figure of 66,000 may be artificially inflated by input from Western Europe, it is safe to conclude that the ocean is the major sink for PCBs.

Polychlorinated dibenzofurans, which occur as impurities in many PCB grades, were identified in the rice oil involved in the Yusho incident, and in the livers of the individuals poisoned by the oil; they have recently been reported at the trace level in the fat of seal and turtle [110].

Polychlorinated Terphenyls

Since the earliest report of the occurrence of PCT in the environment [140], observations have been sporadic. The small amount of data available is presented in Table 8 [78, 123] and Table 9 [5, 24, 38, 41, 57, 67, 78, 111, 129, 135].

It may be concluded that the distribution of PCT is still sporadic and local; PCT and PCB are often found in the same sample, in a ratio varying from 1 in 100 to a maximum of 1 in 20. The only exception is human fat, where the concentration in Japan is often as high as that of PCB.

Polychlorinated Naphthalenes

There are even less data on PCB occurrence than on PCT; the few observations on biological samples are summarised in Tables 8 and 9 [37, 126, 127, 129, 137]. The only reports of distribution around a manufacturing unit [37] suggest low background levels in air, water and soil.

Polybrominated Biphenyls

Almost all the data available are derived from the Michigan cattle food contamination incident in 1973. Levels of up to 200 mg/kg in fat of cattle were recorded on farms where contamination was heavy; dairy products from the area showed levels of 1–15 mg/l. Tissues from human subjects living on or near the affected farms were also analysed; the highest recorded level was 175 mg/kg in the fat of one individual, but most indicated concentrations of the order of 1 mg/kg in body fat. Blood levels ranged from 0.002 to 2.25 mg/l, while milk from nursing mothers was much higher, at 0.2–9.0 mg/l. Since 1974, as a result of a rigid quarantine policy, the tissues of slaughtered cattle rarely exceed the tolerance level of 0.3 mg/kg, and no PBB has been detected in cow's milk [19, 20, 47, 66].

Chlorinated Benzenes

Hexachlorobenzene is frequently reported in analyses of samples for chlorinated pesticides, and occasionally determined in its own right. The data summarised in Tables 8 and 9 [25, 26, 38, 61, 76, 78, 86, 109, 134, 138] indicate widespread distribution in biological tissue, but at very much lower levels than PCB; most of the organisms examined are either fish or birds feeding on aquatic organisms, but HCB has been reported from a range of terrestrial animals at levels up to 10 mg/kg or greater [25]. Occurrence in water, either fresh or marine, is very sporadic and at low concentrations; it is not normally detected in the atmosphere.

The lower chlorinated benzenes (di-, tri-, tetra- and penta-) have been found in sewage and effluents [15, 40, 60, 51, 120] in surface and ground waters in the Rhine Basin [10, 50, 56], and in some biological tissues [69, 134]. The high levels of p-dichlorobenzene shown in Tables 8 and 9 were recorded in Japan, where they are assumed to be caused by its considerable use as an odoriser. Monochlorobenzene is not normally detected in any samples. Both hexachlorobenzene and pentachloronitrobenzene are found in some foods, due to their use as fungicides, but not normally at levels above 0.05 mg/kg.

A large number of chlorinated substituted benzenes, such as chlorotoluenes, chloroanilines, and tetra- and octachlorostyrene, is found is traces in a wide range of sewage and industrial effluents; but they are not normally detected in biological tissues.

Distribution and Degradation in the Environment

Distribution in Air, Water, and Sediments

The compounds discussed in this chapter fall into 2 main groups, in terms of their mobility and distribution behaviour. The first group consists of monochloro- and the dichlorobenzenes, which due to their pattern of use, will be largely released to the atmosphere, where they will be rapidly dispersed; although they have comparatively high vapour pressures, they are also appreciably soluble in water, so are likely to exist in a dynamnic equilibrium between air and water, in which rainout will be matched by evaporation. Factors affecting the balance of this equilibrium are discussed below. Although they adsorb appreciably to active carbon (Table 4) there is little evidence of adsorption to fine particles in the environment.

The major constituents of the second group are PCB, PCT, PCN and hexachlorobenzene, which are characterised by very low solubility in water, low vapour pressures, high octanol-water partition coefficient, and strong tendency to adsorb on solids. As most of the data available on distribution mechanisms relate to PCB, this will be exclusively referred to in the discussion, but the other three products will behave in similar fashion. Tri- and tetra-chlorobenzenes are intermediate between the 2 groups, but as little information is available on their occurrence or behaviour, and the quantities released are small, they will be not further considered.

Assessments of the overall source/transport/sink model of PCBs in the environment [9, 75, 95, 99] all conclude that the most significant factors in the balance are the exchange reactions between water surface and atmosphere (particularly over the oceans), and between water and aquatic sediments.

Volatility is a function of both solubility and vapour pressure, modified by the effect of activity coefficient, and other factors which cause departures from ideal Henry's Law behaviour.

Solubility figures quoted for PCB grades must be arbitrary, as each is a mixture of isomers, whose solubilities range from 5.9 mg/l for 2-chloro-, to 7 µg/l for octachloro- [65]. The quantity in solution therefore represents an equilibrium between the water and the mixture of isomers left as undissolved PCB; this equilibrium takes 100 days or more to be reached [52, 85, 103]. A range of published results is shown in Table 3.

Vapour pressures at 20 °C are very low, and have generally been estimated by extrapolation from curves derived at temperatures of 100 °C or higher. Measurements of volatilization from water, made in the laboratory [89, 103, 128] all show rates higher than would be expected on the basis of calculation from vapour pressure and solubility in water. The actual rates vary, depending on the apparatus used and agitation conditions. PCBs adsorb strongly onto active carbon [23] and are assumed to adsorb equally strongly to both inorganic and organic particles in water, on the basis both of their high octanol-water partition coefficients, and field and laboratory observations [75, 95, 99, 103]. They are also assumed to be present in the oleophilic layer found on the ocean surface, at concentrations much above that in the main body of water, for the same reasons.

All these observations lead to a qualitative view of a Mobile Environmental Reservoir, in which PCBs are mainly discharged to freshwater streams and coastal waters, where they adsorb on particles, and are carried to the oceans; there is significant evaporation from the ocean surface to atmosphere, which acts as a major transport route, but from which PCBs are rained out over a wide land area; material deposited on the land surface is largely re-evaporated, and washed out into rivers. As the physico-chemical properties of the individual isomers vary over orders of magnitude, each process such as evaporation, rainout, adsorption, will tend to classify the products and lead to different congener distributions in different phases. This model has not been quantified. The more permanent sinks consist of product still in service, landfill sites, and stable sediments; the rate of formation of such sediments by deposition from water is not known [27, 95].

Degradation by Chemical Reaction

All the substances under discussion are very stable to hydrolysis, to chemical oxidation, and to thermal decomposition. Degradation by purely abiotic chemical reactions is therefore considered as an unlikely environmental sink [65, 107].

Photodegradation

Experimental studies of the effect of U.V. light, as either natural sunlight or vapour discharge lamps on PCBs in the form of thin films, aqueous dispersions or solutions in organic solvents have shown that direct photolytic decomposition will take place [64, 65, 117]; irradiation of PCB in the vapour phase does not indicate any significant reaction.

Degradation is determined by noting the disappearance of the characteristic peaks of the parent material in GC or MS analysis, and by the appearance of peaks indicating products; the products are not always identified. Reaction pathways include direct dechlorination, and, in the presence of oxygen containing solvents, the formation of substituted hydroxy derivatives; these are illustrated in Fig. 2. They may be accompanied by, or followed by, isomerisation and condensation, leading to the production of terphenyls, tetraphenyls, and dibenzofurans (Fig. 2).

Photodegradation of PCT and PCN has also been demonstrated; but decachlorobiphenyl, and hexachlorbenzene are not significantly photolysed in labo-

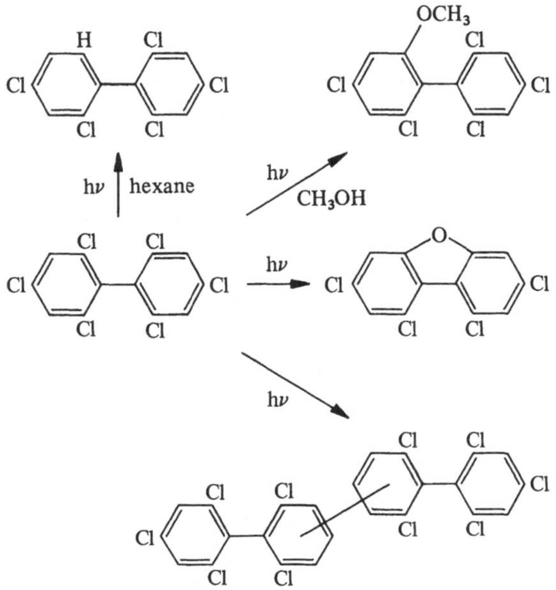

Fig. 2. Photodegradation of 2, 2', 4, 4', 6, 6' hexachlorobiphenyl

ratory studies. Although photodegradation is probably responsible for the disappearance of some chlorinated aromatic compounds in the environment, the distribution of most of the PCBs in the ocean, and on sediments, where the penetration of U.V. light is restricted, makes it unlikely that it is a major sink.

Similar studies with mono- and dichloro-benzenes suggest that even in the atmosphere, neither direct photolysis, nor indirect photo-initiated chemical reactions, play any significant part in degradation [36].

Microbial Biodegradation

The aerobic microbial biodegradation in laboratory studies of the less chlorinated isomers of PCB (5 or fewer chlorine atoms), of some PCTs and PCNs, and of monochlorobenzene is well established [43, 44, 48, 49, 81, 105]; the rates of degradation are reduced by increasing substitution by chlorine, and are also affected by the pattern of distribution of substituents in the molecule. There is no evidence for microbial degradation of higher chlorinated PCB, PCN, or of chlorinated benzenes containing 2 or more chlorines.

Where microbial metabolism occurs, it generally takes the form of a benzene ring-opening reaction, leading to a dicarboxylic acid, or a hydroxy carboxylic acid; a catechol may be an intermediate product. A prerequisite appears to be the absence of substituents on the 2 and 3 positions, thus allowing the formation of an arene oxide or a diol as the first stage of the process. This is schematically illustrated in Fig. 3.

Dechlorination or dehydrochlorination of the benzene nucleus does not occur under aerobic conditions.

Chlorinated anilines are aerobically degraded, by both soil and sewage microorganisms, the monochloro-compounds more readily than the dichloro-derivatives. Chloroanilines are formed in soil as part of the degradation process of the substituted urea herbicides [70]. An alternative pathway is the production of substituted azobenzenes, by dehydrogenation-condensation involving two amino groups.

Many substituted aromatic compounds are also degraded under anaerobic conditions, but the mechanisms are less well understood. There is some evidence that dechlorination can take place; also that chlorinated nitro compounds are converted to amino derivatives which will then follow the above degradation pathways [16].

Bioaccumulation

Examination of the data on occurrence of chlorinated aromatic compounds in the environment, summarised in Tables 8 and 9, makes it clear that PCB, PCT, PCN, and HCB show a marked tendency to accumulate in biological tissues, particularly the fatty tissues of higher animals, both aquatic and terrestrial. Bioconcentration factors of 3–4 orders of magnitude between water and fish may be deduced, with a further 1–2 orders of magnitude between whole fish, and the fat storage tissues of predators on them such as cormorant, heron and seal. Some bioconcentration is also indicated in the case of dichloro- and trichlorobenzenes, but data are not adequate for any quantitative assessment.

There are surprisingly few reports of laboratory bioaccumulation studies with fish or other aquatic organisms, but they confirm the field observations; they also illustrate a strong correlation between the octanol-water partition coefficient and the plateau Bioconcentration Factor (BCF) [8, 17]. This relationship is illustrated in Fig. 4. The BCF for whole fish tissue varies directly with the fat content [124] which can range from 3 to nearly 20%; it is therefore important to report lipid contents when carrying out bioaccumulation studies.

Octanol-water partition coefficients are traditionally determined by shaking the 2 phases together in a separating funnel; this method becomes very inaccurate at log P values above 5, and alternatives based on High Performance Liquid Chromatography or Thin Layer Chromatography have been used [79, 112] to determine accurate values for high molecular weight chlorinated aromatics.

Vertebrate Metabolism

It was recognised at an early stage that the distribution pattern of isomers in vertebrate tissues differed considerably from that in the dominant grades of PCB [18, 65]; the proportion of 5 chlorine isomers and above was much higher than expected. This could be due partly to the physical sorting processes described earlier which would lead to a selectively different exposure pattern in the waters to which the organisms were exposed, partly due to the effect of the rapidly increasing partition coefficient of the more chlorinated isomers, and partly to a higher metabolic

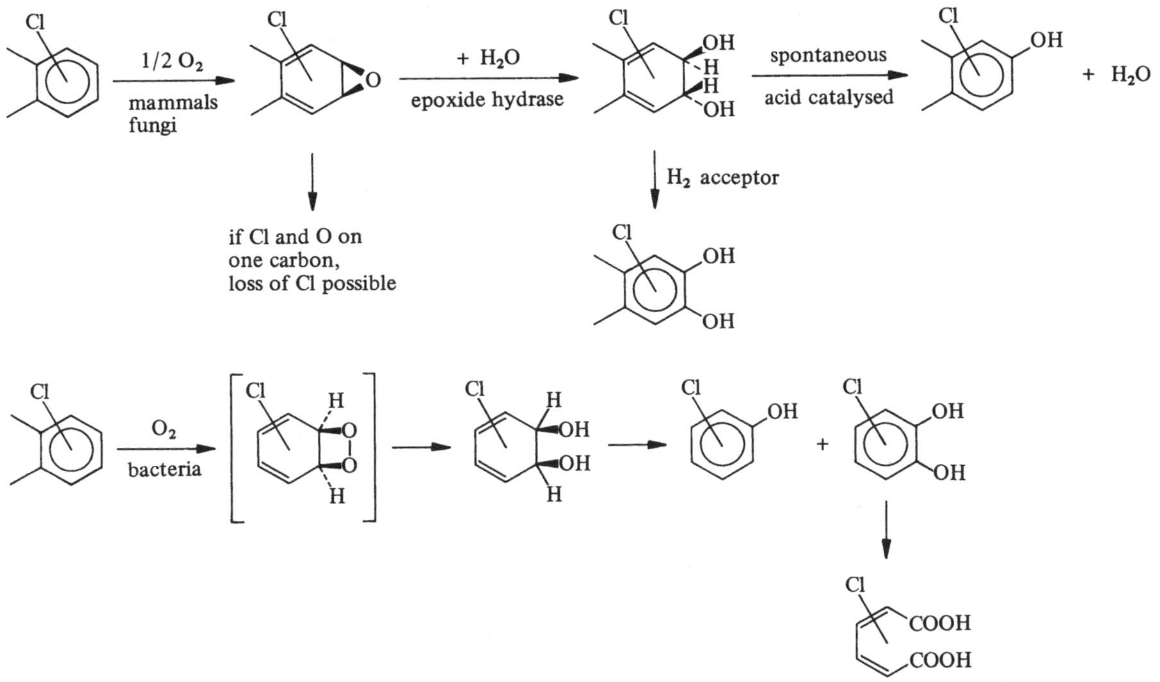

Fig. 3. Reaction sequence for oxidation of (chlorinated) benzene rings by bacteria and higher organisms

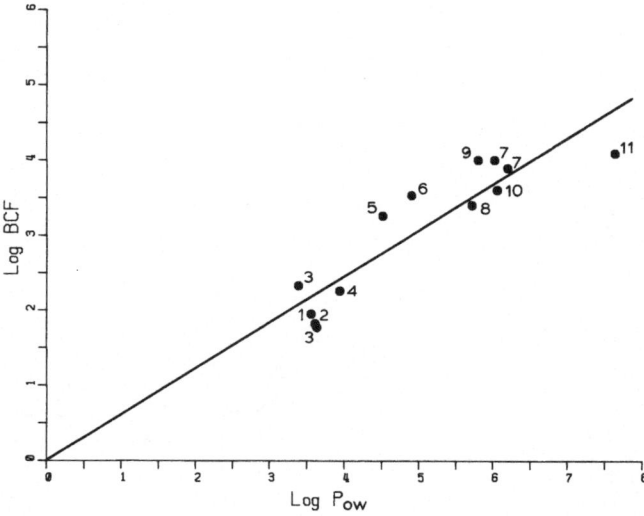

Fig. 4. Relationship between octanol-water partition coefficient (P_{ow}) and bioconcentration factor (Log BCF)

1. 1,2-dichlorobenzene
2. 1,3-dichlorobenzene
3. 1,4-dichlorobenzene
4. 1,2,4-trichlorobenzene
5. 1,2,3,5-tetrachlorobenzene
6. Pentachlorobenzene
7. Hexachlorobenzene
8. 4-chlorobiphenyl
9. 2,3,4,5-tetrachlorobiphenyl
10. 3,4,3',4' tetrachlorobiphenyl
11. 2,4,2',4' tetrachlorobiphenyl

destruction or elimination of the lower isomers by the active centres such as the liver. A number of feeding studies with mammals and birds has confirmed that the latter mechanism, selective metabolism of the isomers containing less than 5 chlorine atoms, is probably the most important [92].

The major route of exposure is ingestion of food containing chlorinated aromatics. All compounds are absorbed by the gut; the rate of absorption is rapidly reduced by increasing halogen content, leading to increased rejection in faeces. Absorbed material is rapidly circulated in the blood stream, from which it is transferred in the short term to liver and kidneys, and in the longer term is laid down in depot fat. Studies with cattle, pigs and chickens show an equilibrium level in fatty tissue which is from 1 to 10 times that in the diet; the higher levels tend to be found in the milk of lactating mothers (including humans), while birds' eggs are at the lower end of the scale.

The behaviour of PCT is similar to that of PCB [68, 118]. Monochlorobenzene and the dichlorobenzenes are sufficiently volatile for inhalation to be a significant route of exposure, at least in the workplace or other indoor atmospheres, they will also be exhaled when exposure ceases. All other compounds are excreted in the urine, after metabolism to hydroxyl compounds, as glucuronide or other conjugates. Metabolism proceeds via an arene oxide (Fig. 3), as a result of mixed function oxidase activity in the liver; compounds containing two adjacent unsubstituted carbon atoms form arene oxides most readily. Chlorobenzenes have been shown to be excreted as the appropriate chlorophenols [125, 134], while the commonest metabolic products of PCBs are the 4- and 2-hydroxy derivatives. Increasing chlorination rapidly reduces the rate of metabolism, but all compounds that can be absorbed are ultimately excreted once exposure ceases [92].

The major route of exposure for fish is the surrounding water, via the gills; excretion probably follows the same route, as fish do not have the ability to oxidise chloro-aromatics to the hydroxyl derivatives to the same extent as mammals [92]. Fish are known to be able to metabolise some substituted naphthalenes [54], but there is no report of chlorinated naphthalenes being in this class. The presence of a simple methyl sulphone derivative of PCB has been reported [73].

Effects on Biological Systems

Mammals

The acute toxicity of all the chlorinated aromatics to mammals is low, with rat LD_{50} values ranging from 500 mg/kg for the dichlorobenzenes, to 10 g/kg and higher for some grades of Arochlor. Many are absorbed through the skin, but do not produce acute systemic effects by this route, except at high concentrations.

Human fatal poisoning incidents have been recorded, particularly for PCB (the Yusho incident, [59]), PCN [81] and hexachlorobenzene [80, 134]. The effects of poisoning are chloracne, hepatic porphyria, liver damage, blood disorders and, at lower concentrations, disorders of the nervous system.

Many of these symptoms have been reproduced in chronic laboratory feeding trials, but only at comparatively high concentrations, of at least 100 mg/kg, rising to 1,000 mg/kg and higher in the diet [17, 115, 134].

Mink and Rhesus monkeys have been shown to be particularly sensitive to PCB, with marked loss of growth, and particularly reproductive failure, at dietary levels as low as 3 mg/kg [77, 93].

Helle [58] reports a correlation between high residue levels in seals and reproductive impairment, caused by uterine abnormalities; but although the blubber of non-pregnant females contained 100 mg/kg PCB, it also contained a similar quantity of DDT.

Birds

The acute toxicity of chlorinated aromatic compounds to birds is low, as in the case of mammals. They are more sensitive in chronic laboratory feeding studies, showing significant reduction in growth, and reduced reproductive performance, at dietary levels of as low as 10 mg/kg [108, 115]. Reproductive failure is not normally correlated with eggshell thinning.

Incidents of poisoning in the wild are difficult to interpret, as the body burden of chlorinated compounds consists of a mixture of pesticides, in addition to any PCB present [62].

Aquatic Organisms

The chlorinated benzenes show appreciable acute toxicity to fish, in laboratory determinations of 48 h to 96 h LC_{50}; reported values range from 0.1 mg/l for p-dichloro-, to 30 mg/l for monochloro-, with values for all other compounds lying in between. Hexachlorobenzene is not reportedly toxic to any species over 96 h [101, 105].

PCBs cause mortality in chronic tests with fish extending over long periods, of 100 days or more, at levels as low as 10 μg/l; but the evidence is not clear. Some sub-acute effects of PCN on crustacea have been reported at 20 μg/l [83], and of other substituted naphthalenes at less than 1 mg/l [102] but the ecological significance of these effects is doubtful.

There is no evidence of toxicity to lower invertebrates, or to microorganisms, except at high levels [33, 101].

Treatment and Disposal of Wastes

Sewage, Liquid Effluent and Water Treatment

All the compounds under discussion occur at low concentrations in sewage, and in many industrial effluents. During the course of normal biological treatment, the volatile compounds (mono- and dichlorobenzenes) are lost by volatilization, while the rest are reduced, by 50%–70%, mainly by adsorption on sludge, with little evidence of biodegradation [40, 60, 119].

Similarly, treatment of river water by sedimentation and flocculation will reduce chloro-aromatic levels by removal on the sludge; the normal process of chlorination for disinfection is unlikely to produce any significant change, other than to slightly increase the percentage of highly chlorinated isomers. Where further reduction is necessary, especially in the treatment of water for potable use, treatment by adsorption on active carbon, or exchange resins, will effectively remove most chlorinated aromatics; those less readily adsorbed will tend to be lost by volatilization. The use of ozone, with or without U.V. light, is not very successful [23, 84, 122].

Incineration or Combustion

Although all the chlorinated aromatics have considerable heat stability, they will partially decompose under conditions of combustion at above about 500 °C; both they, and their decomposition products, will therefore be found in the waste gases from both low-temperature incinerators and open fires, burning waste materials containing them. There is also evidence that aromatic compounds may by synthesised in small quantities by pyrolysis; it is possible that chlorinated aromatics may be so formed, as low molecular weight chlorinated aliphatics such as methyl chloride certainly are (see Chap. on C_1/C_2 halocarbons). The ash resulting from such combustion will also contain at least the more chlorinated isomers of the high boiling compounds such as PCB, as well as reaction products such as PCDF and PCDD [110].

To ensure complete destruction of PCB and similar compounds, it is necessary to use carefully designed incinerators operating at temperatures of 1,200–1,300 °C.

Chemical Treatment of Solid Wastes or Slurries

A process has been proposed which uses chemical dechlorination, in which all the chlorine is removed as sodium chloride; few details are available.

Disposal to Landfill Sites

The only effective alternative to incineration for solid wastes is disposal in controlled landfill sites; wastes containing significant amounts of chlorinated aromatics are residues from manufacture or processing, small items such as capacitors, waste heat transfer or hydraulic fluids, sludges from sewage or effluent treatment, and increasingly, materials such as food, feeding stuffs, dyes, etc, which contain higher than acceptable contaminant levels.

Laboratory studies suggest that there will be little movement of chlorinated aromatics from landfill sites which are protected from excessive run-off, or from percolation through cracks and fissures; evaporation will be negligible, except for any exposed surface, while leached material will adsorb firmly to the soil immediately below the landfill site [39, 52, 53, 95].

If extra security is required, wastes may be immobilised by one of the proprietary inorganic gel processes.

A large quantity of PCB has accumulated in the sediments at the bottom of industrial estuaries such as the Hudson, in USA, to such high levels that it is acting as a continuous input to the Mobile Environmental Reservoir. The spoil resulting from controlled dredging of these estuaries can be deposited on carefully chosen sites, which enable the supernatant wastes to be drained off and settled to remove the fine suspended solids, which contain much of the PCB [139].

Implications of Regulatory Control Measures

Society may judge, on the basis of available evidence, that restrictions may need to be placed on the continuing uncontrolled production, use and disposal of some chemicals; it has decided so very clearly in the case of PCB, and to a lesser extent in the case of some other chlorinated aromatics. Such action may be taken on a completely voluntary basis by the manufacturer(s), by agreement with a national regulatory agency or by imposed statutory control.

The range of options open include:
1. Complete ban on manufacture
2. Control or restriction on uses
3. Control on processing/storage/transport and distribution
4. Limits or consent procedures on waste disposal
5. Quality or residue tolerance standards in products/foods/water/air.

In the case of PCB, the first actions were taken on a voluntary basis by Monsanto, the major manufacturer; in 1970 they suspended sales for all except enclosed uses, and from 1976 they effectively ceased manufacture.

In 1973, the OECD member states took the decision to restrict uses to enclosed or semi-enclosed, and to regulate manufacture, distribution and waste disposal. In 1976, the Toxic Substances Control Act in the USA provided the framework for a complete ban on manufacture, and for strict controls on disposal of used PCB. Similar controls are in force in Japan and the EEC member states.

The implications of such a ban are twofold:

1. Replacement materials may need to be generated [13]. In the case of PCB, the need for stability, fire resistance, and good electrical characteristics has restricted the search to compounds that may still have the same properties of persistence and bioaccumulation, that led to the concern about PCB in the first place; principal contenders are alkylated biphenyls and terphenyls, silicones, phosphate esters and high molecular weight phthalates.

2. There is a large stock of product still in service which must eventually be disposed of, without adding to the environmental burden. In the case of PCB, there are also a number of landfill sites, and areas of marine sediment rich in waste PCB, which could also add to the Mobile Environmental Reservoir. In the USA, EPA has set 50 mg/kg as the PCB level above which a material is designed as falling within the regulations covering labelling, containers, transport, and waste disposal; this includes a very large quantity which must be kept in controlled storage, and eventually either incinerated, or destroyed by an equally acceptable method.

The result of these controls is a pattern of distribution within the environment which has been described in summary form in this chapter; if any fall in average concentrations is occurring, data are inadequate to quantify any trend.

A voluntary cessation of production of PBB, falling production of PCT and PCN, reduced use of hexachlorobenzene as a fungicide, and effective control of the waste disposal of heavy end residues containing HCB, have reduced environmental input to low levels.

The other type of regulatory measure is that which sets tolerance limits, especially on food and water. Although there are few specific limits for potable water supplies, the generally accepted target of 1 µg/l for chlorinated pesticides is rarely exceeded by any of the chloro-aromatics, and concentrations are generally much lower.

In the US, which has the most extensive food tolerance and action limits, acceptable levels for PCB range from 0.2 mg/kg for infant foods, through 3 mg/kg for meat, to 5 mg/kg for fish and shellfish. Rules are proposed to reduce the action level for fish to 2 mg/kg, but this would reject a significant proportion of the present edible fish catch.

Such wide-ranging control measures are not applied to any of the other chloroaromatic compounds.

References

1. Ahnoff, M., et al.: Water Res. *13*, 1233 (1979)
2. Anliker, R.: Swiss Chem. J. No. 1–2 (1981)
3. Anon.: PCB Conference – Stockholm 1970. National Swedish Environment Protection Board (1970)
4. Anon.: PCB Conference II – Stockholm 1972 National Swedish Environment Protection Board (1973)
5. Anon.: Annual Report of the Government Chemist – London (1976)
6. Anon.: National Conference on PCB Chicago 1975. EPA Washington (1976)
7. Ballschmiter K., Zell, M.: Int. J. Environ. Anal. Chem. *8*, 15 (1980)
8. Barrows, M. E., et al.: Bioconcentration and elimination of selected water pollutants by Bluegill Sunfish. Dynamics, Exposure and Hazard Assessment of Toxic Chemical. (Ed. R. Haque) Ann Arbor, 1980
9. Bidleman, T. F.: High Molecular Weight Chlorinated Hydrocarbons in the air and sea. Rates and Mechanisms of Air-Sea Transfer. Workshop on Tropospheric Transport of Pollutants to the Ocean. 79, 1975
10. Borneff, J., Hartmetz, G., Fischer, A.: Die Belastung von Oberflächenwasser mit Chlorobenzolen. Spez. Ber. Kernforschungsanlage Jülich 45. Organohalogen-Verbindungen Umwelt 51 (1979)
11. Brinkmann, U. A. Th., et al.: J. Chromatog. *129*, 193 (1976)
12. Brinkman, U. A. Th., De Kok, A.: J. Chromatography *129*, 451 (1976)
13. Broadhurst, M. G.: Environmental Health Perspectives N2, 81 (1972)
14. Burger, E. J.: Polychlorinated Biphenyls – A Case Study. NRC/EPA Washington (1974)
15. Callahan, M. A., Ehreth, D. J., Levins, P. L.: Proc. Natl. Conf. Munic. Sludge Manage. *8*, 55 (1979)
16. Chacko, C. I., Lockwood, J. L., Zabik, M.: Science *154*, 893 (1966)
17. Chiou, C. T. et al.: J. Envir. Sci. Techn. *11*, 475 (1977)
18. Cook, J. W.: Environmental Health Perspectives *1*, 3 (1972)
19. Cordle, F.: Environmental Health Perspectives. *24*, 157 (1978)
20. Cordle, F., Kolbye, A. C.: Vet. Hum. Toxicol. *20*, 245 (1978)
21. Dept. of Environment, UK: Methods for the Examination of Waters. Organochlorine Insecticides and PCB. HMSO (1978)
22. DHEW: Health Effects of PCBs and PBBs. Summary and Conclusions. Environmental Health Perspectives *24*, 191 (1978)
23. Dobbs, R. A., Cohen, J. M.: Carbon Adsorption Isotherms for Toxic Organics. EPA 600/8-80-023 (1980)
24. Doguchi, M.: Ecotoxicology & Environmental Safety *1*, 239 (1977)
25. Drescher-Kaden, U.: Occurrence of Organohalogen Compounds in Terrestrial Wildlife. Spez. Ber. Kernforschungsanlage Juelich 45. Organohalogen-Verbindungen Umwelt (1979)
26. Dubrawski, R., Falandysz, J.: Marine Pollution Bull. *11*, 15 (1980)
27. Eisenreich, S. J., Hollod, G. J., Johnson, T. C.: Env. Sci. & Technol. *13*, 569 (1979)
28. EPA: Prioritized Guidelines for Environmental Fate Testing of Chlorobenzene. EPA-506/5-77-001. EPA (1977)
29. EPA: Sampling and Analysis Procedures for Screening of Industrial Effluents for Priority Pollutants. EPA (1977)
30. EPA: Chlorinated Naphthalenes – Ambient Water Quality Criteria. EPA (1978)
31. EPA: Polybrominated Biphenyls – Status Assessment of Toxic Chemicals. EPA (1979)
32. EPA: Hexachlorobenzene – Status Assessment of Toxic Chemicals. EPA (1979)
33. EPA: Master Scheme for the Analysis of Organic Compounds in Water Interim Protocols. EPA (1980)
34. EPA: Dichlorobenzenes – Ambient Water Quality Criteria. EPA (1980)
35. EPA: Polychlorinated Byphenyls. Ambient Water Quality Criteria. EPA (1980)
36. EPA: Fate of Toxic and Hazardous Materials in the Air Environment. EPA-600/53-80-084 (1980)
37. Erickson, M. D., et al.: J. Assoc. Official Analytical Chemists *61*, 1335 (1978)
38. Falandysz, J.: Mar. Poll. Bull. *11*, 75 (1980)
39. Farmer W. J., et al.: J. Soil So. Soc. Am. *44*, 676 (1980)

40. Feiler, H.D., Vernick, A.S., Storch, P.J.: Fate of Priority Pollutants in POTW's. Proc. Natl. Conf. Munic. Sludge Manage. *8*, 72 (1979)
41. Fukano, S., Doguchi, M.: Bull. Environmental Contamination and Toxicology *17*, N5 613 (1977)
42. Fuller, B.: Environmental Assessment of PCBs in the Atmosphere. EPA-450/3-77/045 (1976)
43. Furukawa, K., Matsumura, F.: J. Agric. Food. Chem. *24*, 251 (1976)
44. Furukawa, K., Tomizuka, N., Kamibayashi, A.: J. Pharmacobio-Dyn. *3*, S-4 (1980)
45. Gaffney, P.E.: Science *183*, 367 (1974)
46. Gaskin, D.E., Holdrinet, M., Frank, R.: Arch. Environm. Contam. Toxicol. *7*, 505 (1978)
47. Getty, S.M., Rickert, D.E., Trapp, A.L.: Critical Rev. in Environmental Control *7*, 309 (1977)
48. Gibson, D.T.: Microbial transformation of aromatic pollutants. In: Aquatic Pollutants – Transformation and Biological Effects, Pergamon *1*, 187, 1977
49. Gibson, D.T.: Microbial metabolism. In: Handbook of Environmental Chemistry, Springer-Verlag *2A*, 161, 1980
50. Giger, W., Molnar-Kubica, E., Wakeham, S.: Volatile chlorinated hydrocarbons in ground and lake waters. Aquatic Pollutants: Transformation and Biological effects. Pergamon *1*, 1977
51. Glaze, W.H., Henderson, J.E., Smith, G.: Analysis of New Chlorinated Organic Compounds in Municipal Wastewaters After Terminal Chlorination. Identification and Analysis of Organic Pollutants in Water. Ann. Arber Science 1976
52. Griffin, R.A., Chian, E.S.K.: Attenuation of Water Soluble Polychlorinated Biphenyls by Earth Materials. Environmental Geology Notes 86, Illinois State Geological Survey 1979
53. Griffin, R.A., Au, A.K., Chian, E.S.K.: Mobility of PCB and Dicamba in Soil Materials. Proc. Natl. Conf. Munic. Sludge Manage. *8*, 183 (1979)
54. Gruger, H.E.H., et al.: Aquatic Toxicology *1*, 37 (1981)
55. Harms, U., Drescher, H.E., Huschenbeth, E.: Meeresforschg. *26*, 153 (1977/78)
56. Hartmetz, G., Fischer, A.: Vorkommen von Chlorbenzolen im Rhein und seinen Zuflüssen. Wa-Bohu-Ber. *3*, 86 (1978)
57. Hassell, K.D., Holmes, D.C.: Bull. Environmental Contamination and Toxicology. *17*, 618 (1977)
58. Helle, E.: Ambio *5*, 261 (1976)
59. Higuchi, K. (Ed.): PCB poisoning and pollution. Academic, Tokyo 1976
60. Hites, R.A.: Sources and Fates of Industrial Organic Chemicals. Proc. Natl. Conf. Munic Sludge Manage. *8*, 107 (1979)
61. Holden, A.V.: Organochlorine residues in blubber of grey seals from the Farne Islands. ICES CM 1978/E:41 (1978)
62. Howard, E.B., Esra, G.N., Young, D.: Acute Foodborne Pesticide Toxicity in Cormorante and Seagulls. In: Animals as Monitors of Environmental Pollutants NAS (1979)
63. Howard, P.H.: Preliminary Hazard Assessment of Chlorinated Naphthalenes, etc. EPA 560/2/74-001 (1973)
64. Hutzinger, O., Safe, S., Zitko, V.: Environmental Health Perspectives. *1*, 15 (1972)
65. Hutzinger, O., Safe, S., Zitko, V. (Eds.): The Chemistry of PCB's. CRC (1974)
66. Isleib, D.R., Whitehead, G.L.: Trace Subst. Environ. Health *9*, 47 (1975)
67. Jan, J., Josipovic, D.: Chemosphere *11*, 863 (1978)
68. Jan, J., Malnersic, S.: Bull. Env. Contam. Toxicol. *19*, 772 (1978)
69. Jan, J., Malnersio, S.: Bull. Env. Contam. Toxicol. *24*, 824 (1980)
70. Janicke, W., Hilge, G.: GWF-Wasser/Abwasser *121*, 131 (1980)
71. Jensen, G.E., Clausen, J.: J. Toxicol. Environmental Health. *5*, 617 (1979)
72. Jensen, S., et al.: Nature *224*, 247 (1979)
73. Jensen, S., Jansson, B.: Ambio *5*, 257 (1976)
74. Jolley, R.L., et al.: Determination of chlorination effects on organic constituents in natural and process waters. In: Identification and analysis of organic pollutants in water. Ann Arbor 1977
75. Kalmaz, E.V., Kalmaz, G.D.: Ecological Modelling *6*, 223 (1979)
76. Kerkhoff, M., De Boer, J.: Organochlorine residues in a harbour porpoise found dead in the Dutch Wadden Sea in 1971. ICES CM 1977/N:2 (1977)
77. Kimbrough, R., et al.: Environmental Health Perspectives. Animal Toxicology *24*, 173 (1978)
78. Koeman, J.H., Stasse-Wolthuis, M.: Environmental Toxicology of Chlorinated Hydrocarbon Compounds in the Marine Environment of Europe. EEC (1978)
79. Könemann, H., et al.: J. Chromatog. *178*, 559 (1979)

80. Koss, G., et al.: On the Effects of the Metabolites of Hexachlorobenzene. Chem. Perphyria in Man, Elsevier 1979
81. Kover, F.D.: Environmental Hazard Assessment Report – Chlorinated Naphthaleses. EPA Washington (1975)
82. Landner, L., Skoglund, P.O.: Polyklorenade Bifenyler I Vattenmiljo. IVL Stockholm (1977)
83. Laughlin, R.B., Neff, J.M.: Marine Environmental Res. 2, 275 (1979)
84. Lawrence, J., Tosine, H.M.: Environmental Science and Technology 10, 381 (1976)
85. Lee, M.C., Chian, E.S.K., Griffin, R.A.: Water Research 13, 1249 (1979)
86. Leoni, V., D'arca, S.U.: Science of the Total Environment 5, 253 (1976)
87. Lindskog, A.: Atmospheric Transport of PCB and HCB. Inst. Vatten-Luftvardsforsk, B 527, IVL (1980)
88. Luckas, B., Wetzel, H., Rechlin, O.: Die Nahrung 24, 405 (1980)
89. Mackay, D., Wolkoff, A.W.: Env. Sci. and Technol. 7, 611 (1973)
90. Macleod, K.E.: Sources of emissions of polychlorinated biphenyls into the ambient atmosphere and indoor air. EPA-600/4-79/022 (1979)
91. Margeson, J.H.: Methodology for measurement of polychlorinated biphenyls in ambient air and stationary sources – A review. EPA-600/4-77/021 (1977)
92. Matthews, H., et al.: Environmental Health Perspectives 24, 147 (1978)
93. McNulty, W.P.: Primate News 12, 34 (1974)
94. Miyata, H., Kashimoto, T., Kunita, N.: J. Food Hygiene Soc. Japan 18, 126 (1978)
95. NAS: Polychlorinated Biphenyls. Washington 1979
96. Nelson, N.: Environmental Res. 5, 249 (1972)
97. Neufeld, M.L., Sittenfield, M., Walk, K.F.: Polybrominated Biphenyls – Market Input Output Studies. EPA 568/6-77/817
98. Nicholson, M.L., Moore, J.A. (eds.): Health Effects of Halogenated Aromatic Hydrocarbons. Ann. N.Y. Acad. Sci. 320 (1979)
99. Nisbet, I.C.T., Sarofim, A.F.: Environmental Health Perspectives. 1, 21 (1972)
100. OECD: Report on Protection of the Environment by Control of PCB. OECD Paris 1979
101. Oshida, P.S.: Toxicity of a chlorinated benzene to sea urchin embryos. S. California Coastal Water Res. Proj. 187 (1977)
102. Ott, F.S., Harris, R.P., O'Hara, S.C.M.: Marine Envir. Res. 1, 49 (1978)
103. Paris, D.F., Steen, W.C., Baughman, G.L.: Chemosphere 7, 319 (1978)
104. Peakall, D.B.: Critical Rev. Environ. Control. 5, 469 (1975)
105. Pearson, C.R.: Unpublished
106. Phillips, D.J.H.: Envir. Pollut. 16, 167 (1978)
107. Pomerantz, I., et al.: Environmental Health Perspectives 24, 133 (1978)
108. Prestt, I., Jefferies, D.J., Moore, N.W.: Environ. Pollut 1, 3 (1970)
109. Pucetti, G., Leoni, V.: Marine Pollution Bull. 11, 22 (1980)
110. Rappe, C., et al.: Nature 292, 524 (1981)
111. Renberg, L., Sundström, G., Reutiergardh, L.: Chemosphere 6, 477 (1978)
112. Renberg, L., Sundström, G., Sundh-Nygard, K.: Chemosphere 9, 683 (1980)
113. Risebrough, R.W., et al.: Nature 220, 1098 (1968)
114. Risebrough, R.W., Lappe, B.D.E.: Environmental Health Prospectives 1, 39 (1972)
115. Roberts, J.R., et al.: Polychlorinated Biphenyls: Biological criteria for an assessment of their effects on environmental quality. National Research Council Canada. NRCC 16077 (1978)
116. Rostval, L., Szokolay, A., Uhnak, J.: Die Nahrung 24, 359 (1980)
117. Safe, S., et al.: Photodecomposition of Halogenated Aromatic Compounds. Identification and Analysis of Organic Pollutants in Water. Ann Arbor Science 1976
118. Sekita, H., et al.: J. Hygienic Chemistry 21, 307 (1975)
119. Shannon, E.E., Ludwig, F.J., Valdmanis, I.: Polychlorinated Biphenyls in municipal wastewaters: An assessment of the problem in the Canadian Lower Great Lakes. Environmental Protection Service, Canada. Research program. Project No. 73-3-8 (1976)
120. Sheldon, L.S., Hites, R.A.: Env. Sci. and Technol. 13, 574 (1979)
121. Siuda, J.F., et al.: Gas Chromatographic detection of naturally occurring halogenated compounds derived from marine organisms. Food-Drugs from the Sea Proceedings 1972. Marine Technol. Soc. 1972

122. Sonneborn, H. M. (Ed.): Gesundheitliche Probleme der Wasserchlorierung und Bewertung der dabei gebildeten halogenierten organischen Verbindungen. Wabolu-Ber. *3* (1978)
123. Stratton, C. L., Sosebee, J. B.: Env. Science and Technology *10*, 1229 (1976)
124. Sugiura, K., et al.: Chemosphere *6*, 359 (1979)
125. Szokolay, A., Uhnak, J., Madaric, A.: Nahrung *24*, 381 (1980)
126. Takeshita, R., Yoshida, H.: Studies on Environmental Contamination by Polychlorinated Naphthalenes – III Human Body. Eisei Kagaku *25*, 24 (1979)
127. Takeshita, R., Yoshida, H.: Studies on Environmental Contamination by Polychlorinated Naphthalenes IV. Marine Fish. Eisei Kagaku *25*, 29 (1979)
128. Tofflemire, T., Shen, T. T.: Volatilization of PCB from Sediment and Water. Experimental and Field Data. Mid-Atl. Ind. Waste Conf. *11*, 100 (1979)
129. Vannucchi, C., Sivieri, S., Ceccanti, M.: Chemosphere *1*, 483 (1978)
130. Vermeer, K., Peakall, D. B.: Marine Pollution Bull. *8*, 205 (1977)
131. Verschueren, K.: Handbook of Environmental Data on Organic Chemicals. Van Nostrand 1977
132. Wassermann, M., et al.: Ann N.Y. Acad. Sci. *320*, 69 (1979)
133. Westin, R.A.: Polychlorinated Biphenyls 1929-79. EPA/560/6-79/004 (1979)
134. WHO: Toxicological Appraisal of Halogenated Aromatic Compounds Following Groundwater Pollution. WHO (1980)
135. Wright, L. H., et al.: J. Anal. Toxicol. *2*, 76 (1978)
136. Yamada, T.: Polychlorinated Biphenyls in Organs and Tissues of Japanese 1972–1974. Nara Igako Zasshi *30*, 445 (1979)
137. Yoshida, H., Takeshita, R.: Studies on Environmental Contamination by Polychlorinated Naphthalenes VI. Sea Water. Eisei Kagaku *25*, 334 (1979)
138. Young, D. R., Heesen, T. C.: Chlorinated Benzenes in Palos Verdes flatfish. S. California Coastal Water Res. Proj. 149 (1977)
139. Zimmie, T. F., Tofflemire, T. J.: Disposal of PCB Contaminated Sediments in the Hudson River. Proc. Conf. Geotechnical Practice For Disposal of Solid Waste Matter. ASCE (1978)
140. Zitko, V., Hutzinger, O., Choi, P. M. K.: Environmental Health Perspectives *1*, 47 (1972)
141. Zitko, V.: Potentially Persistent Industrial Organic Chemicals Other than PCB in: Ecological Toxicology Research Plenum 1975

Volatile Aromatics

E. Merian

International Association of Environmental Analytical Chemistry and
Swiss Association for Environmental Research
CH-4106 Therwil, Switzerland

M. Zander

Rütgerswerke AG
D-4620 Castrop-Rauxel, Federal Republic of Germany

Benzene, toluene, the xylenes, ethyl benzene, styrene, isopropyl benzene (cumene), chlorobenzene, the dichlorobenzenes, nitrobenzene, diphenyl, naphthalene, 1-methyl-naphthalene, 2-methyl-naphthalene, 2,6-dimethyl-naphthalene, and 2,6-diisopropyl naphthalene are the most important volatile aromatics of commercial significance. These compounds do not occur in the environment naturally except as constituents of petroleum which may seep into the oceans from underground deposits. The occurrence of these products in the environment is thus mainly or exclusively of man-made origin, as distinct from other hydrocarbons such as methane, terpenoids, polycyclic aromatics.

Anthropogenic volatile aromatics exist and enter in the environment both as ingredients of oil fractions and as pure raw materials. For instance the content of toluene and xylenes in motor spirit is much higher than the amounts produced in isolated form. When interpreting figures, one has to keep this in mind, and it should be clearly investigated what the figures mean (including avoiding double counting). For the interpretation of concentration figures the following conversion factors for the atmosphere can be used: 1 ppm (volume) benzene = 2,7 ppm (weight) = 3,2 mg/m^3, 1 ppm (volume) toluene = 3,2 ppm (weight) = 3,8 mg/m^3, 1 ppm (volume) xylene/ethylbenzene = 3,7 ppm (weight) = 4,4 mg/m^3.

Production (Source, Use, Shipment) and Emissions

Volatile Aromatics from Coal and Lignite

Coke-oven gas and coal tar are formed in the carbonization of coal to metallurgical coke [1]. Referred to dry coal, 0.7%–1.5% benzene hydrocarbons can be condensed and washed out of the coke-oven gas [1]. Of these, three-quarters is benzene and the rest are toluene, xylenes, and industrial-grade benzenes [1]. On a worldwide basis, about 16 million t p.a. coal tar are produced by the coal carbonization process, of which about 75% are processed in coal tar refineries (USA 3.5 million t, western europa 4.4 million t) [1, 1a]. In 1965, the USA produced 435,000 t benzene, 23,000 t toluene, 23,000 t xylenes and 220,000 t naphthalene from coal [2, 3] but now the figures are somewhat lower (for example: 165,000 t naphthalene in 1978 [3a].

Lignite can be carbonized in the absence of air, the products obtained being coke, tar, gas, and water (water of decomposition) from the low-temperature carbonization process [1]. In 1965, the FRG[1] produced 1.3 million t tar this way, from which some 50,000 t carburettor fuel and about 400,000 t diesel fuel were obtained [1].

Volatile Aromatics from Petroleum

In 1978, world production of petroleum amounted to 3 056 million t [4]. Crude oil contains between 16.5% (South Louisiana) and 21.9% (Kuwait) aromatics; benzene and its homologues account for 3.9%–4.8%, and naphthalene and its homologues for 0.7%–1.3% [5]. In the order of magnitude the annually extracted petroleum before refining contains about 5 million t benzene, about 12 million t toluene, about 22 million t xylenes and about 30 million t naphthalenes.

It is estimated that annually some 6 million t crude oil and refinery products are introduced into the sea [6]. Of this amount, probably about 40% is crude oil (natural influences, in-shore production, tanker accidents and transportation by tanker) and about 60% are further processed petroleum fractions (refineries, rivers, town sewage, transportation means, precipitation from the atmosphere) [6]. This annual emission into the seas can, in addition to other fractions, correspond to about 75,000 t benzene, 480,000 t toluene, 560,000 t xylenes, and at least 40,000 t naphthalenes.

In 1969, the USA produced 42% gasoline, 22% gas/diesel oil, 7% kerosene/jet fuel and 7% heavy fuel oil from crude oil, while western Europe produced 18% gasoline, 27% gas/diesel oil, 4% kerosene/jet fuel and 37% heavy fuel oil from crude oil – besides other products (including internal refinery consumption) [1]. World demand in 1977 totalled between 530 million t (excluding eastern Europe) [7] and 580 million t [8] motor fuel (carburettor fuel), of which the USA accounted for 291 million t, the FRG 19 million t, and Switzerland 2.6 million t [7, 4]. Straight-run gasoline contains about 8% volatile aromatics, catalytically cracked gasoline about 30%, and reformed gasoline about 64% [7], mainly toluene and C_8-aromatics and small quantities of benzene. Gas and diesel oil can consist of about 33% naphthenes, 27% paraffins, 25% aromatics (C_9 and higher aromatics), and 15% isoparaffins with a boiling point of 150 to 400 °C [5]. The consumption of petrochemical raw materials amounted to 64 million t in Western Europe in 1977, 16% or 10 million t of which were aromatics [9].

---

1 FRG = Federal Republic of Germany

Every year, about 50 million t hydrocarbons are discharged into the atmosphere from refineries and automobile exhausts, about 20 million t from incineration plants, and about 20 million t from other man-made sources [10, 11], of which only part is aromatic. These emissions could contain some 10–20 million t volatile aromatics [12]. Another estimate is based on world emissions of hydrocarbons from stationary plants of 54 million t (USA 7–18 million t) and from mobile sources (transportation means) of 34 million t (USA 12–20 million t) [13]. It has also been calculated that world evaporation losses of hydrocarbons from the production and processing of petroleum amount to 44.7–68 million t (approx. one-quarter of this in refineries), and emissions from the combustion of by-products in the oil processing industry to 28 million t [5]. In these figures, too, the volatile aromatics naturally account for only a portion, perhaps one-fifth.

Motor Fuel (Carburettor Fuel)

The world annual requirement of about 550 million t motor fuel is produced from selected mineral oil fractions. These are further processed and if necessary combined, [1, 7] depending on their original properties, since the demands made on the various grades – particularly with regards to anti-knock rating – can vary considerably. For example, the USA consumes 69% regular grade gasoline, 11% non-leaded gasoline and 20% premium grade gasoline [10], while in the FRG the situation in 1975 was 41% regular grade and 59% premium grade gasoline, and in Switzerland in 1975 17% regular grade and 83% premium grade gasoline [10].

In the USA in 1978, non-leaded gasoline contained 1.2% benzene, regular grade 1.2% benzene, and premium grade 1.1% benzene [14]. The total aromatics content in American gasoline amounts to 30.2% for non-leaded, 27.7% for regular grade, and 39.0% for premium grade gasoline [14].

In the FRG, premium grade gasoline contains 35%–55% aromatics [8], with an average of 3.6% benzene [15], 10%–16% toluene, 12%–18% C_8-aromatics, and 17%–20% higher aromatics [16]. In Switzerland, premium grade gasoline contains on average 35% aromatics, with 2.7% benzene. Regular grade gasoline in the FRG contains 20%–30% aromatics [8], with an average of 2.8% benzene [15], 4%–12% toluene, 8%–13% C_8-aromatics, and 15%–18% higher aromatics [16]. The Swiss regular grade gasoline contains an average of 24.5% aromatics, with 1.8% benzene [10]. According to other studies, motor spirits contain 38.5%–53.9% aromatics, with 0.2%–15% benzene, 7.3%–18.4% toluene, 11.5%–18.0% C_8-aromatics, and 11.2%–20.3% higher aromatics [5]. Between 1970 and 1972, the average content of benzene in carburettor fuels in western Europe amounted to some 4% (premium grade about 6%, regular grade about 2.5%) with isolated cases of extremes up to 16% [17]. On the assumption that about one-quarter premium grade and three-quarters regular grade gasoline is consumed on a worldwide basis, the annual consumption rate of 550 million t motor fuel could contain approximately 10–15 million t benzene, 30 million t toluene, 40 million t C_8-aromatics, and 53 million t higher aromatics, which are burnt and lost as valuable raw materials.

Attempts have been made to estimate the quantities of motor fuel that evaporate into the atmosphere during storage, transloading and transportation. A total emission factor of 0.44% (averaged over the whole year) has been calculated for this in the FRG [16], which would correspond to total emissions of some 100,000 t per annum. In Switzerland, annual losses by evaporation are estimated at 20,000 t [10]. Another assessment estimates worldwide losses by evaporation at 6 million t [8]. The composition of these emissions has also been the subject of careful investigation in the FRG [16]. Referred to the hydrocarbon content, the emissions contain *inter alia* 0.5%–2.8% benzene, 0.2%–1.8% toluene, and very small quantities of higher aromatics (presumably because the vapour pressure is much lower) [16]. No allowance was made for possible losses in the motor vehicles. On the basis of these data, it can be deduced that worldwide roughly an additional 100,000 t benzene, 50,000 t toluene, and 15,000 t higher aromatics are discharged into the atmosphere through evaporation of motor fuel.

An additional load on the environment is imposed by the unburnt portion in automobile exhaust gases. In 1969, US motor vehicles emitted 16.9 million t hydrocarbons [13], which contained about 2.4% benzene (about 400,000 t), about 3.1%–7.9% toluene (about 800,000 t), and about 1.9% m- and p-xylene (about 300,000 t). In the FRG, total hydrocarbon emissions from motor vehicles were estimated at some 2.7 million t [12], while in Switzerland estimates range from 130,000 t [12] to 31,700 t [10] to 41,282 t [18]. In this context, it must be borne in mind that the exhaust gases from the combustion of motor fuel with a 50% aromatics content contain about 20% aromatics and about 25% olefines [19]. This agrees with the observation that immission data in urban areas have approx. 80% aliphatics and approx. 20% aromatics ([12], see also later). In the gasoline engine combustion process, a portion of the higher benzene homologues undergoes thermal dealkylation as a function of the combustion mode [13, 20–23]. Therefore – and because the vapour pressure and the stability of benzene are higher [24] – the concentration of benzene in relation to the concentration of volatile aromatics is higher in the atmosphere, than in the motor spirit. In the FTP-Emission-Cycle vehicles with Otto-engines emit about 7 mg benzene/km, about 13 mg toluene/km and about 3 mg benzaldehyde/km, whereas those with Diesel-engines emit about 6 mg benzene/km, about 1 mg toluene/km and about 2 mg benzaldehyde/km [24]. It is estimated that the automobile exhaust gases totalling some 34 million t p.a. worldwide contain about 1 million t benzene, about 2 million t toluene, and about 1 million t higher aromatics.

Diesel Fuel and Gas Oil

In the USA, production of diesel oil (as fuel) and gas oil (for heating) – which are practically identical – amounts to about half that of gasoline, i.e. about 150 million t [1]. The 1974 consumption in the FRG was some 80 million t, 10 million t of which in the form of diesel fuel [25, 26]. Switzerland has an annual consumption of about 580,000 t diesel fuel and a total of some 9 million t gas oil [18, 26]. It can therefore be assumed that annually on a worldwide basis some 300 million t of this product are used for motor vehicles and a multiple of this figure for heating purposes. Since the relevant petroleum fractions are not sufficient to meet demand, heavy oils are additionally converted by hydrocracking into gas/diesel oil [1]. Diesel fuel can also be produced from coal by hydrogenation [1]. Diesel oil has a boiling range of 150 °–400 °C and, in addition to naphthenes, paraffins and isoparaffins, also contains approx. one-quarter aromatics [5]. These are in particular aromatics higher than C_8 (alkyl benzenes and alkyl naphthalenes). Since western Europe alone had a consumption of 402.6 million t gas/diesel oil in 1974 for heating and powering vehicles [26], it can be expected that annually about 100 million t higher volatile aromatics are burnt in these countries alone. On a worldwide basis, this quantity could be three or four times that figure.

Emissions from the consumption of diesel fuel are low, however. There is virtually no evaporation. In 1969, hydrocarbon emissions from diesel-powered vehicles in the USA amounted to 200,000 t (as compared with 16.9 million t from gasoline engined vehicles) [13]. Of this figure, the C_{10}–C_{24}-hydrocarbons account for a relatively low portion, but are nevertheless responsible for the offensive odour [13]. A few hundred different components have been found, some are components of the diesel oil but others are formed by combustion [13]. The more important constituents are indanes, tetralines, alkyl benzenes, naphthalenes, indoles, acenaphthenes, and benzothiophenes [13]. It appears that the quantitatively less important aromatic aldehydes and ketones are responsible for the odour [13]. At a rough estimate, the total world emissions of alkyl benzenes and naphthalenes must amount to some 200,000 t. Diesel particulate emissions contain neutral fractions with microbial mutagenic activity [26 a].

Benzene

Benzene consumption in the USA was 3.8 million t [3] in 1970 and 6.4 million t [27] in 1976. In the FRG, demand has risen from 0.8 million t [3] in 1970 to 1.1–1.2 million t [28] in 1979. In 1977, the western world had capacities for the production of isolated benzene of 18.1 million t, including 7.0 million t in the USA and 6.5 million t in western Europe [9]. On a worldwide basis, the production of pure benzene is probably in excess of 15 million t. A smaller proportion is obtained from coal [3] – about one-quarter in the FRG [2], less in the USA. Three-quarters of the production in the FRG comes from petroleum, which is subjected to catalytic reforming and pyrolysis, or from which the alkyl benzene fractions are dealkylated [28]. In the USA benzene is also produced from olefines [27]. These figures do not include consumption of benzene in motor fuel, which is in the same order of magnitude, i.e. 10 million t.

A further potential source of environmental pollution is the burning of wood, garbage and organic wastes – such as gardening refuse (where about 1.4 kg benzene are obtained from one tonne) and rice waste [13, 29, 30]. Since the combustion gases from wood total only about 400,000 t hydrocarbons [13] per annum in the USA, this benzene source is of minor significance in relation to the 26–30 million t consumed as raw material or in gasoline.

With regard to its application, of the 6.4 million t benzene produced in the USA, 3.2 million t are used for the production of ethyl benzene/styrene, 1.2 million t for the production of cumene, 1 million t for the production of cyclohexane, 340,000 t for the production of nitrobenzene, 330,000 t for the production of alkyl benzenes, 150,000 t for the production of chlorobenzene, and 120,000 t for the production of maleic anhydride [27]. It is a disquieting thought that in 1968 50,000 t benzene were also used as solvent in the USA. In western Europe it is expected that the 1979 benzene requirement will have been 3.4 million t for ethyl benzene/ styrene, 1 million t for phenol, 900,000 t for cyclohexane, 500,000 t for chloro- and nitrobenzene, 300,00 t for maleic anhydride, and 200,000 t for alkyl benzenes [3]. Certain environmental consequences may result from the fact that in all processing stages pipes and storage vessels for benzene must be heated, for example using gly- col as heat carrier, because of the relatively high melting point of + 5.5 °C [3].

With regard to losses, it was estimated in the USA that in 1977 20,000 t benzene were discharged annually by the chemical industry from 63% of the identified emission sources [27]. Of this, benzene production accounts for 7,300 t and the pro- cessing to maleic anhydride for 8,000 t [27]. The rest is divided among other appli- cations in the chemical industry. A total of about 40% escape from storage tanks [27]. On the basis of these figures, it can be calculated that losses at present amount to about 0.17% of the quantities processed [27]. A reduction of these 20,000 t to 5,000 t would cost the chemical industry in the USA about 78 million $ in invest- ments and about 56 million $ in additional annual operating expenses [27]. On a worldwide basis, benzene losses in the chemical industry probably amount to some 100,000–200,000 t. To this figure must be added losses during production, trans- portation and processing of petroleum, transportation and distributing of motor fuel and combustion in gasoline engines in the order of 3 million t benzene. As al- ready mentioned a great part of the benzene in the motor fuel is in fact burnt in the combustion process, but, at the same time, there is an additional amount of benzene produced through dealkylation, which is emited [13, 20–23]. As mention- ed, the total annual load imposed on the sea is probably some 75,000 t benzene as part of petroleum fractions.

Toluene

In the USA, 1.6 million t toluene were produced and consumed in 1970 [3]. This quantity increased to 3.4 million t by 1978 [30a]. The FRG produced 189,000 t toluene in 1970 [3]. The European Community is consuming 1.0 million t in 1980 [30a]. World production is estimated at more than 5 million t, or about one third of the benzene figure. However, as already mentioned, an additional 30 million t toluene are consumed annually as a constituent of motor spirit. Toluene is ob- tained to a lesser extent from coal – by fractional distillation of light tar oil [1, 31] or as by-product from metallurgical coke production [32] – and to a greater extent from petrochemical operations [31, 32]. Toluene is also a by-product in the manu- facture of styrene [31].

Toluene is used as an intermediate for synthesis processes or as a solvent, or it is dealkylated to benzene [32]. In the USA, 240,000 t were used as solvent in 1968 [13]. In Japan, toluene also plays an important role as a solvent in paint and print- ing ink works, and annual emissions of 250,000 t [33] to 600,000 t [34] toluene are estimated from this application alone. In addition to emissions in workplaces or the environment, accidents are also possible in connection with the use of toluene and other solvents. For example, a leakage of toluene and other chemical products in July 1979 caused a major fish kill in the Main/Rhine rivers [35].

Losses, therefore, can be estimated worldwide to be in the order of 1–1.5 million t from the use as solvent. To this must be added toluene losses associated with the production, transportation and processing of petroleum in the order of 3–4 million t as a portion of the total hydrocarbon losses, and some 2 million t toluene as a component of automobile exhaust gases. In these figures toluene emissions are included which are either produced through dealkylation of higher volatile aromatics [13, 21–23] in combustion processes or through burning of wood and garbage [13, 29, 30]. Compared with these emissions, the losses during transportation and distribution of gasoline (estimated worldwide at about 50,000 t toluene) and the losss in the chemical industry of about 50,000–100,000 t toluene (assumed to be half as much as benzene) are relatively modest. As mentioned, the total annual load imposed on the sea is probably some 500,000 t toluene as part of petroleum fractions.

Xylenes, Ethyl Benzene and Styrene

The USA produced 1.6 million t xylenes [3] and 2 million t ethyl benzene/styrene [36] in 1970, and 2.9 million t xylenes and 3.8 million t ethylbenzene in 1978 [30 a]. The FRG produced 219,000 t xylenes [3] and some 300,000 t ethyl benzene/styrene [36] in 1970. For western Europe, production of these raw materials in 1980 is expected to be three times the 1970 figure [3]. In the USA in 1978, 0.5 million t o-xylene and 1.6 million t p-xylene were produced individually. The capacities in the European Community in 1980 were 0.6 million t o-xylene and 0.8 million t p-xylene [30 a]. Accordingly, world production of C_8-aromatics will probably amount to about 20 million t. However – as already mentioned – an additional 40 million t are consumed annually as a constituent of motor fuel.

As with toluene, xylenes can be obtained by fractional distillation from light tar oil [1, 31], but are primarily obtained by cracking operations from petroleum fractions [1, 31]. Ethyl benzene is also obtained simultaneously, although it is mainly synthesized from benzene. Xylene can also be produced by disproportionation of toluene alongside benzene [37]. Several industrial processes are available for separation of the C_8-aromatics [37].

Over 90% of ethyl benzene is obtained by catalytic alkylation of benzene with ethylene [36, 38]. It is converted to styrene by dehydrogenation [36, 38].

Xylenes are used *inter alia* as solvents, for example in the USA 180,000 t p.a. [13] and in Japan 90,000 t [33], particularly in the paint and printing ink industries (see also [5]). In eastern Europe it is also used for numerous applications in the shoe industry. o-xylene (besides naphthalene) is mainly oxidized to phthalic anhydride from which dyestuffs, phthalodinitrile and – by esterification with alcohols – plasticizers and raw materials for paints and varnishes are obtained [36, 38]. p-xylene is oxidized and processed with methanol to terephthalic acid dimethyl ester, from which polyesters are made [36, 38, 39]. For example, in 1972 2.9 million t polyester fibres were produced worldwide [39] and in 1979 more than 5 million t, which corresponds to 2.8 million t of p-xylene [30 a]. Polystyrene and copolymers (including expanded plastics) in particular are polymerized from styrene, and the world annual production of these products around 1975 was about 5 million t [39, 40], which confirms that about 5 million t ethyl benzene/styrene are also consumed. With regard to emissions, it must be borne in mind that toluene and other alkyl benzenes also evaporate during the manufacture and polymerization of styrene, and the storage of polystyrene products; even though they are not significant in terms of quantity, they can cause offensive odours [38, 41, 42].

The xylene losses introduced into the environment are probably made up of some 0.5 million t from solvent losses, about 2 million t from losses during the production, transportation and processing of petroleum as a portion of the total hydrocarbon losses, and some 0.5–1 million t as a component of automobile exhaust gases. Compared with these emissions, the losses during transportation and distribution of gasoline (estimated worldwide at 10,000 t) and the losses in the chemical industry (estimated at less than 50,000 t) are relatively low, also because the vapour pressure of the C_8-aromatics is lower than that of benzene and toluene. As mentioned, the total annual load imposed on the sea is probably some 600,000 t xylenes as part of petroleum fractions.

Chlorinated Benzenes and Nitrobenzene

In the USA, 150,000 t benzene were used in 1976 for the production of chlorobenzene [27]. It is assumed in western Europe that 500,000 t benzene will have been used in 1979 as initial product for chloro- and nitrobenzene [3]. Based on these figures, the annual world production of chlorinated benzenes can be estimated at 600,000–800,000 t. Direct chlorination of benzene with chlorine gas is effected on a continuous basis in the presence of catalysts, such as aluminium, mercury, iron, sulphur chlorides or molybdenum chloride [32, 36, 43], following which the mixture is separated and purified by washing, chemical treatment and distillation operations. Chlorobenzene is used mainly for the production of phenol, chloronitrobenzenes, DDT, and as a solvent. o-dichlorobenzene is a solvent, whereas p-dichlorobenzene is used as an intermediate or (in addition to naphthalene) as a mothproofing agent [292].

Chlorinated benzenes may also result from volatile aromatics in water during the disinfection process with chlorine (see later sections).

In the USA, 340,000 t benzene were used for the production of nitrobenzene in 1976. Consumption of benzene for aniline production in the USA was 208,000 t in 1975, and is estimated at 281,000 t in 1980 [3]. In western Europe, it is assumed that 500,000 t benzene will be used in 1979 as the initial product for chloro- and nitrobenzene [3]. On the basis of these figures, it can be deduced that the annual world production of nitrobenzene is in the order of 800,000 to 1 million t. Nitrobenzene is generally obtained on a continuous basis by the reaction of benzene with a mixture of sulphuric acid and nitric acid [36], and it is used in particular as the initial product for intermediates – particularly aniline – and as a solvent.

The losses of chlorinated benzenes and nitrobenzene in the chemical industry are probably low – also because of the low vapour pressure – and will total less than 20,000 t. Losses from its use as a solvent could be higher, perhaps in the order of 50,000–100,000 t worldwide. The quantities of chlorinated aromatics formed during the chlorine disinfection of waters are difficult to assess, but are probably low.

Higher Alkylated Benzenes, Naphthalene, Tetraline, and Alkylated Naphthalenes

The total annual production of petroleum probably contains some 30 million t naphthalene [4, 5]. The world production of naphthalene from coal tar and from petroleum (excluding the Eastern Bloc countries) is estimated at 1 million t [36]. In contrast, as estimated earlier, the quantities of higher aromatics present in gas/ diesel oil are much higher. Annually, these could amount to 100 million t [5, 26] in western Europe alone, and 300–400 million t alkylated benzenes (with more than 9 C-atoms) and – possibly alkylated – naphthalenes and tetralines worldwide.

Most of the naphthalene produced is oxidized to phthalic anhydride [2, 36]. In addition, naphthalene is used as the basis for other intermediates – primarily as initial products for dyestuffs – [36], or it is used besides and with p-dichlorobenzene as mothproofing agent and insecticide. Products similar to diesel oil – such as Aerotox 3470 – together with alkylated benzenes and naphthalenes are also used as solvents for pesticides, and are then introduced into the environment by spraying [44]. 1,1-phenyl-xylyl-ethane and diisopropyl naphthalene are PCB substitutes [45].

Worldwide, the losses of higher alkylated benzenes and – possibly alkylated – naphthalenes in the chemical industry are estimated to be low, particularly also because of the low vapour pressure of these compounds, and are probably less than 10,000 t. Their use as mothproofing agent and solvent for pesticides could be of greater significance, possibly hundreds of thousands of tonnes of these compounds are discharged into the environment. Even more critical are the losses from petroleum fractions not used for chemical purposes. Total emissions associated with the production and consumption of gas/diesel oil could be in the order of 200,000 t alkylated benzenes and naphthalenes. As mentioned, the total annual contamination of the seas by naphthalene as a part of petroleum fractions is probably at least 40,000 t. Since pollution of the seas by naphthalenes is indeed a problem – they are particularly water-insoluble and therefore accumulate in organisms – it has been the subject of intensive studies [46–48, 49–51]. Losses of alkylated benzenes and – possibly alkylated – naphthalenes are probably critical, because of their low vapour pressure and higher chemical stability they tend more to accumulate in water and organisms (see later sections).

Analytical Methods

Gas-liquid chromatography is the method of choice, in most cases, for determining volatile aromatics in environmental samples. Generally, the samples contain numerous other components, aliphatic hydrocarbons in particular, in addition to volatile aromatics, so that the separation capacity of the gas-liquid chromatographic system has to meet high demands, and since analysis must nearly always be carried out in the trace range, the same applies to sensitivity. Gas-liquid chromatography with glass capillary columns using the flame ionization detector best meets such demands [52].

If the samples contain relatively small portions of volatile aromatics and large quantities of aliphatic material, determination of the aromatics can be substantially improved by using selective stationary phases [53] and/or selective detectors. The electron capture detector is particularly suitable for the selective detection of aromatics alongside aliphatic or alicyclic compounds, since it is highly sensitive to aromatics and has a low sensitivity to saturated hydrocarbons [54].

The number of possible isomers increases with the higher C number, but more difficult separation problems associated with the volatile aromatics themselves do not arise until the range of naphthalene homologues is attained. More frequently, selective stationary phases have been proposed for the gas-liquid chromatographic separation of the isomeric dimethyl naphthalenes. Nematic crystal phases [in particular 4-(2,2-ethoxy-ethoxy-carbethoxyoxy)-cinnamic acid ester of the 4-hydroxy-4'-methoxy azobenzene] [55] or calcium chloride [56] are interesting suggestions. Calcium chloride is used with chromosorb as the support, and operations are carried out under conditions of gas-solid chromatography in packed columns.

Because of the fundamental disadvantages of chromatographic methods in the qualitative analysis, samples of complex composition are frequently analysed by a combination of the chromatographic method (as the separation method) and spectroscopic methods (as identification method). The on-line coupling of gas-liquid chromatography in high efficiency columns with mass spectroscopy [57] is, in most cases, the best suited method also for the analysis of environmental samples containing volatile aromatics. But there are also interesting alternatives, for example: the recently developed on-line coupling of gas-liquid chromatography with laser-excited multi-channel Raman spectroscopy [58].

Environmental analytical methods must frequently satisfy a number of conditions:
(1) The continuous monitoring of pollution by volatile aromatics, for example of workplaces, industrial agglomerations or densely populated areas with a high traffic density, makes it essential that the methods can be automated, which in turn requires a low susceptibility to failure.
(2) Where application of the methods requires preparation of the samples, this should involve the least possible working stages in order to prevent material losses and minimize the time required for the analysis. These conditions, too, are best met by gas-liquid chromatography.

In a recent publication [59], the sensitivity to aromatics of the photo-ionization detector, which is greater by 1–2 orders of magnitude as compared with the flame ionization detector, was used for the direct (i.e. without preconcentration by trapping or freeze-concentration) and automated gas-liquid chromatographic determination of benzene and benzene homologues in ambient air. At a signal/noise ratio of at least 2:1, the lower limit of detection for benzene, toluene and m-xylene was 0.3 ppb.

Even though the direct gas-liquid chromatographic determination (i.e. without preconcentration) of volatile aromatics in water samples is possible in principle, the detection limits are high – not better than about 1 mg/l. Techniques that are considerably more sensitive are those in which a flow of gas (e.g. helium) is passed through the water sample. The gas picks up the volatile aromatics, then flows through a drying tube (e.g. magnesium perchlorate), and finally to a cooled column containing an adsorbens (e.g. activated alumina). This is then placed in the gas chromatography carrier gas flow and heated to desorb the aromatics [60]. This method can be used for benzene, for example, with a sensitivity in the order of 10^{-1} µg/l.

If certain compounds have to be detected in complex environmental samples with high selectivity and sensitivity, it is occasionally possible to make use of their different UV absorption spectra. An example is the detection of naphthalene/naphthalene homologues alongside benzene, benzene homologues, alkanes and alkenes, all of which absorb at shorter wavelengths than do naphthalene/naphthalene homologues. While conventional UV-spectroscopy is less suited for this task in most cases derivative UV-spectroscopy is an improvement because of its better signal/noise ratio and consequently higher sensitivity, but fluorescence and phosphorescence spectroscopy with selective excitation are the most suitable methods [61]. Detection limits in the ppm range can be attained with optimized techniques. Generally it is difficult to distinguish between basic aromatic systems (for example: naphthalene) and their homologues using electron spectroscopic methods. This is a restrictive factor for the determination of individual components in mixtures, but is an advantage for "group analyses." Methods which permit the "total luminescence" defined as the function I (λ_{ex}, λ_{em}), where I is the intensity and λ_{ex} and λ_{em} are all possible excitation and monitoring (fluorescence and phosphorescence) wavelengths, to be measured as sample-characteristic contours or isometric projections would seem to be particularly suitable for group analysis by luminescence spectroscopy [62].

New developments in this area can be expected from methods based on spectroscopy using lasers. For example, the spatial coherency characteristic of laser irradiation, which permits beam control and focussing over greater distances, was used in an interesting approach to measure the air pollution rate over an industrial site and its surrounding area. The method requires no sampling and permits the recording of changes in air pollution concentrations in terms of area and time [63].

In view of the importance of determining volatile aromatics in the environment several reviews have been written:
- Development of Method for Carcinogenic Vapor Analysis in Ambient Atmospheres (EPA-650/2-74-12) July 1974
- Methods for the Determination of Hydrocarbons in the Atmosphere, CONCA-WE-Report (The Hague, NL) Nr. 3, Febr. 1976
- The Measurement of Carcinogenic Vapors in Ambient Atmospheres (EPA-600/7-77-055), June 1977
- Analysis of Organic Air Pollutants by Gas Chromatography and Mass Spectroscopy (EPA), June 1977
- Organic Characterization of Aerosols and Vapor Phase Compounds in Urban Atmospheres (EPA-600/3-78-031), March 1978
- Determination of Benzene from Stationary Sources (EPA-Method 110), May 1978
- Several Bulletins (740 D, 743 E, 769 A, 773 A) about the Determination and Separation of Hydrocarbons, SUPELCO, Inc., Bellefonte, Pennssylvania 16823, USA.

Many other publications give also analytical information, when discussing concentrations in the environment and effects (such as [13, 16, 17, 64, 67, 68, 110, 132, 134, 135, 137, 145, 163] and [195]).

Transport Behaviour in the Environment

While in general a fair amount is known about the transport of petroleum fractions and hydrocarbons, the specific behaviour of volatile aromatics has not been studied to any great extent. It is known that they are carried through the air over relatively long distances. For example, aromatic fractions from automobile exhaust gases in the Zurich area are found in Lake Zurich and the river Glatt [64, 65]. Similarly, ethyl benzene, xylenes, n-propyl benzene, ethyl toluenes and trimethyl toluenes from Sydney have been detected in remote regions of New South Wales [66].

Only insufficient information is available concerning the exchange of volatile aromatics between the atmosphere and water. It can be reliably assumed that rainfall precipitates volatile aromatics from the atmosphere, whereupon they are introduced directly or indirectly into surface water. The occurrence of aromatic fractions from automobile exhaust gases in Lake Zurich and the river Glatt, mentioned above, provides evidence that such an exchange actually does take place [64, 65] for other water ways see [67, 68] and a later section [see also 291, 292].

There are indications that gasoline fractions from oil spillages disappear from the sea completely within six hours by evaporation [69]; this in contrast to higher boiling fractions. Of the approx. 6 million t petroleum fractions introduced into the sea per annum, approx. 600,000 t originate from natural sources [5, 6], approx. 600,000 t are precipitated from the atmosphere and some 1.9 million t originate from rivers. However, the proportion of volatile fractions is not known [6]. Volatile aromatics may perhaps account for about one-fifth of these quantities or – in relation to their vapour pressure – less. Near the surface of the sea one finds 1–10 ppb of natural hydrocarbons, less in deeper waters [6]. Organic pollutants may increase the absorption of hydrocarbons in water [13]. Aromatic hydrocarbons are the relatively best water soluble components of crude oil and petroleum fractions. Therefore they have an increased mobility in waters.

Various attempts have been made to develop models for the behaviour of hydrocarbons and their derivatives in the environment. For example, on the basis of tests with fuel oil, Nigerian crude oil and Venezuelan crude oil, a numerical model was developed [70] to permit the forecasting of changes in oil characteristics, specific density, and the oil portion remaining after accidents in waters – taking into account evaporation and weather factors, particularly the effects of temperature and wind. Another model [71] was used to study the physico-chemical and biological behaviour of ten relatively simple benzene derivatives (including chlorobenzene, nitrobenzene and hexachlorobenzene). In principle, it should be possible to estimate potential environmental contamination from the knowledge of the basic molecular properties of the pollutants, such as water solubility, partition coefficient for lipid/water, reactivity as determined by electron density and from fundamental linear free energy values as Hansch's π and Hammett's σ. The former provides indications concerning bioaccumulation, the latter concerning possible reactivity. Radioactive substances were used to determine partition coefficients for lipid/water (chlorobenzene 150, nitrobenzene 62, hexachlorobenzene, 13,560), ecological magnification (chlorobenzene 650, nitrobenzene 29, hexachlorobenzene 1,166) and biodegradability indices (chlorobenzene 0.014, nitrobenzene 0.023,

Table 1. Physical constants important for predicting environmental behaviour [2, 31, 32, 37, 38, 50, 72–77]

	Melting point °C	Boiling point °C	Flash point °C	Density 20 °C	Saturated vapour pressure at 20 °C	Water solubility at 20 °C g/m³	Partition coefficient octanol/water log K_{ow}
Benzene	+ 5.5	80.1	− 11	0.88	74.3 mm	1,780	2.13
Toluene	− 95	110.6	+ 4	0.87	22.4 mm	515	2.69
o-xylene	− 25.2	114	+ 17	0.88	5.2 mm	175	3.15
m-xylene	− 4.8	139	+ 23	0.87	6.4 mm	162	
p-xylene	+ 13.2	138	+ 25	0.86	6.8 mm	185	
Ethyl benzene	− 95	136	+ 15	0.87	7.1 mm	152	3.15
Styrene	− 33	146	+ 34	0.91	Approx. 250		
Isopropyl benzene	− 96	153	+ 36	0.86	4 mm	50	3.43
Diphenyl	+ 70.4	255		0.99		7.0	4.03
Chlorobenzene	− 45	132	+ 28	1.11	8.8 mm	472	2.84
1,4-dichlorobenzene	+ 53	174	+ 65	1.53	0.5 mm	87.2	3.38
Nitrobenzene	+ 5.7	210.8	+106	1.20	0.27 mm	205.7	1.85
Naphthalene*)	+ 80.2	211	+ 79	1.14	0.05 mm	31.7	3.37
1-methyl-naphthalene*)	− 22	242		1.02		28.5	3.86
2-methyl-naphthalene	+ 37	241		1.01		25.4	4.11
2,6-dimethyl-naphthalene*)	+111	262		1.00		2.0	4.38
2,6-diisopropyl-naphthalene		279		0.97		Approx. 0.1	

* See also J. M. Neff "Polycyclic Aromatic Hydrocarbons in the Aquatic Environment", Tables 31/32, Applied Science Publishers 1979

hexachlorobenzene 0.377), and also the partition and degradation in algae and aquatic fauna. As a rule, mainly polar substances are formed by decomposition, such as chlorophenols from chlorobenzene, aniline, nitrophenol, and aminophenol from nitrobenzene, and pentachlorophenol and non-extractable compounds from hexachlorobenzene.

Table 1 shows the physical parameters of volatile aromatics important for predicting environmental behaviour. For some half-live values see [291, 292].

Chemical and Photochemical Reactions

Volatile aromatics are starting materials for numerous commercial synthesis processes. In contrast to the other compounds, benzene is relatively stable [43]. Above 650 °C, diphenyl is formed thermally [43]. It is relatively resistant to oxygen. At 400 °–500 °C and 2–3 bar pressure, with a residence time of 0.1 s, it can be oxidized with large quantities of excess air on catalysts (V_2O_5 on Al_2O_3) to maleic anhydride [43]. Benzene can be relatively easily hydrogenated to cyclohexane [43].

In the environment, in the absence of reactive impurities, the only reactions which can occur are oxidations of side chains and possibly dechlorination with light of wavelengths over 290 nm [5]. The photochemical reactivity is enhanced by nitric oxides or solid particles, on whose surfaces catalytic reactions take place [5]. For example, it has been proved that toluene and xylenes decompose on silica gel under the effect of ultraviolet radiation within four days (with a wavelength over 230 nm) [78]. Toluene, xylenes and other alkyl benzenes are auto-oxidized in the atmosphere by oxygen roughly at the same rate as ethylene, especially in the presence of catalyzing air impurities [13].

In the USSR and in Japan processes were developed to keep the working atmosphere – for instance of the paint and printing industry – free from volatile aromatics. The processes are based on the catalytic oxidation or decomposition of toluene and other offensive and malodorous components [79–81]. The polluted air is oxidised catalytically at 80 °–100 °C at with calgon activated coal coated with metals [79] or at platinum catalysts on alumina oxide [80]. The components may also be decomposed at 225 °–300 °C quantitatively by cobalt oxides on a carrier, such as aluminum oxide [81]. Replacing the volatile aromatics or preventing them from evaporating into the air, are better solutions of the problem in this case.

The Photochemical Decomposition of Benzene

The reaction rate of photochemical decomposition at room temperature in the presence of oxygen and nitric oxides is lower for benzene than for other hydrocarbons [82]. In photochemical reactions, acrolein and glyoxal [83, 84] and primarily nitrobenzene and nitrophenols [85, 86] are formed from benzene in the presence of nitric oxides, with o- and p-nitrophenol being formed from nitrobenzene as intermediate [87]. If benzene is exposed to light in the presence of carbon monoxide and nitric oxides, then p-nitrophenol, 2,6-dinitrophenol and 2,4-dinitrophenol are formed [88].

The Photochemical Decomposition of Toluene and Xylenes

Reaction rates in the presence of oxygen and nitric oxides [82], in the presence of hydroxyl radicals [89] or in the presence of mixtures of nitric oxides of varying composition [90] have been compared for various aromatics. m-xylene and trimethyl benzenes react substantially faster than the other xylenes, toluene, and in particular benzene. In the presence of nitric oxides, the following products are formed from aromatics with side chains particularly from toluene: methyl glyoxal, dimethyl glyoxal, and biacetyl [83, 84], larger quantities of m-nitrotoluene, 2-methyl-4-nitrophenol, 2-methyl-4,6-dinitrophenol, benzaldehyde, benzyl nitrate, o-cresol, and p-nitrophenol [91–93].

Reactions Taking Place in Automobile Emissions, and Smog Formation

During the combustion process in gasoline engines, benzene, toluene, acetylene, acrolein, and aromatic aldehydes [94], of which the latter have a particularly high photochemical activity, are formed from volatile aromatics by cracking and disproportionation, and also by oxidation.

Methyl nitrate, acrolein, aceton, and other products are formed as a result of ring cleavage in the presence of nitric oxides [95]. When automobile exhaust gases are exposed to ultraviolet light in the presence of nitric oxides, 30% of the benzene, 68% of the toluene, and 84% of the xylene decompose within six hours [96].

Aromatics with side chains – and aldehydes formed from them – contribute towards the formation of photochemical smog [12, 13]. Reactions with both ozone and nitric oxides have been studied [13]; in the latter case, ozone, peroxyacetyl, nitrate, hydroxyl radicals, and other active oxidants are formed which, among other things, irritate the eye. In proportion to the oxygen concentration, aldehydes, ketones, esters, and carboxylic acids, which either occur as aerosols or agglomerate, are formed from volatile aromatics when exposed to ultraviolet light [97]. Aerosols occur preferentially in the presence of both nitric oxides and sulphur dioxide [97]. Investigations into correlations between the formation of products and the concentration of hydrocarbons and nitric oxides revealed that a maximum of aldehydes occurs after about one to two hours, and that 70%–80% of the o-xylene and toluene react in the presence of light [98]. Apart from sulphur dioxide, water, ammonia, ozone or 2-methyl-2-butene can also catalyze the reactions and so promote aerosol formation [99].

Degradative Reactions of Aromatics in Water [see also 291]

Several studies have been made concerning the degradation and decomposition of aromatic-containing oils in the sea following tanker collisions. The aromatic content is no more than 20%, but it has a poor viscosity/temperature behaviour, is relatively toxic, and leads to sticky tars as a consequence of oxidation [100]. Some aromatic hydrocarbons, however, are fairly soluble in water (see Table 1) and can therefore relatively easily be transformed or decomposed by chemical or biologic reactions [101]. More critical are naphthalene and its methylated derivatives [101]. Sunlight and/or active oxygen can form products, which are chemically and/or biologically more active [102]. For example, hydroperoxides, phenols, and ketones formed from tetraline, indane, fluorene, and acenaphthenes can be toxic for algae, micro-organisms, invertebrates and fish [102, 103]. Toluene and xylenes can be oxidized with ozone in water. This, however, is much more difficult in the case of benzene, since the oxidation of benzene follows a different reaction pathway [104].

Biodegradation, Effects on Microorganisms and Plants

Microorganisms can degrade aromatic hydrocarbons mainly in water and soils; it is even possible to cultivate sandy soils by the controlled use of certain fertilizers and degradable oil fractions [5]. Certain micro-organisms are capable of living and multiplying in petroleum fractions by degrading the hydrocarbons [5], especially under aerobic conditions. In addition to the decomposition of oil fractions in the sea by microorganisms and other forms of life, there are also other cleaning and distribution mechanisms whereby, for example, plankton can ingest oil droplets and excrete them in a modified physical form [5, 105]. Algae and plants can absorb

10–200 ppm hydrocarbons [5, 106]. The behaviour of the green alga *Oedogonium cardiacum* has been studied in an aquatic ecosystem model [71].

The concentrations of volatile aromatics which occur in the environment are generally – in opposition to the concentrations of olefinic compounds – too low to be harmful for plants [5, 13]. In concentrations of more than 2% – as they may be found when sprayed as solvents for pesticides – volatile aromatics are, however, more ecotoxic than olefines and paraffines [13, 107]. Relatively critical are oxidants, which may be produced from alkylated benzenes in a similar way as from olefinic compounds [5, 13].

Biodegradation of Benzene and Benzene Homologues

Degradation using *Pseudomonas putida* has been relatively well studied. While mammals use only one atom of the necessary oxygen and convert aromatic bonds into an arene oxide, from which a trans-diol is formed by addition of water, bacteria use both atoms of one oxygen molecule to form a dioxetan, which is subsequently reduced to a cis-diol [108–110]. Suitable bacteria must therefore be capable of combining not only oxygen but also hydrogen for the decomposition process [108–110]. The aromatic ortho-diphenols are then formed from the diols by an enzymatic reoxidation which can be further degraded to aliphatic carboxylic acids (Fig. 1).

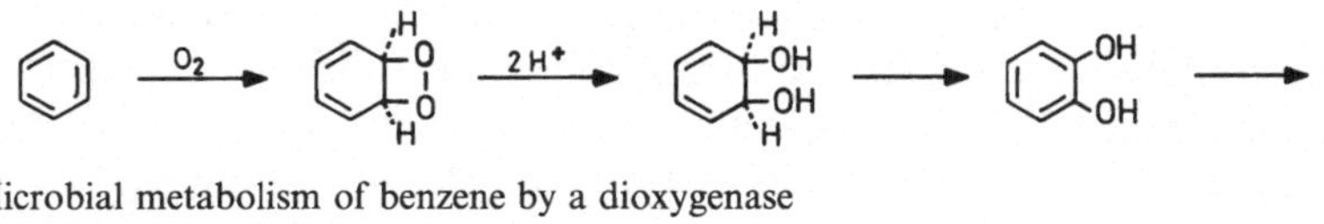

Fig. 1. Microbial metabolism of benzene by a dioxygenase

Some kinds of bacteria can oxidise aromatic compounds when yet another substrate is present [108]. In the presence of hexadecane, for example, Nocardia co-oxidizes ethyl benzene to phenylacetic acid and p-xylene to α,α'-dimethyl muconic acid [108] (Fig. 2).

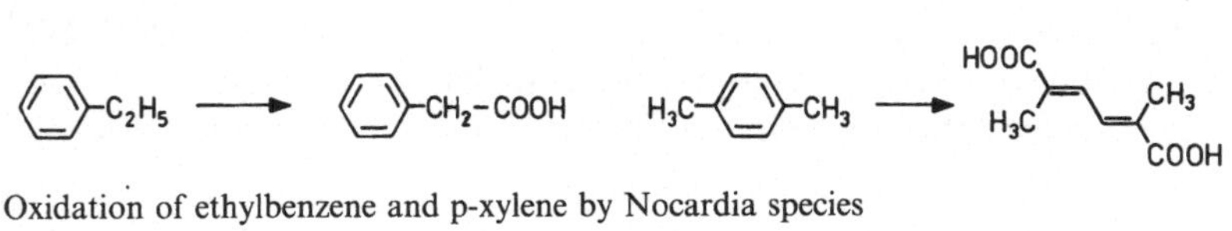

Fig. 2. Oxidation of ethylbenzene and p-xylene by Nocardia species

Xylenes can also be degraded microbiologically to methylbenzoic acids and to xylenols [111, 112]. Since benzoic acids, phenols, xylenols etc. are hydrophilic decomposition products of volatile aromatics, it may be mentioned that these derivatives are easily biodegraded further by microbes [113–120]. Appart from these aerobic biodegredation reactions, also anaerobic biodegradation reactions are possible under special circumstances [113].

The oxidation of cumene and p-cresol in natural waters has also been the subject of chemical and biological (enzymatic) studies. Here, radicals, alcohols and dimers are formed [121]. Recently, attempts have been made to carry out such oxidation/reduction reactions, also on an industrial scale, using biocatalysts – enzymes or immobilized cells [122]. At a half life of the biocatalysts of 3–4 days, it is possible to decompose benzene at 30 °C to carbon dioxide [122].

Biodegradation of Naphthalene and Naphthalene Homologues

Because of their volatile nature, benzene hydrocarbons accumulate relatively little in organisms, especially since some of them have also a relatively high water solubility. Naphthalenes are more critical, and they also play a significant role in oil spillages. The biodegradation of oil fractions has therefore been studied in detail [108, 123]. *Pseudomonas putida* metabolizes naphthalene like benzene to 1,2-dihydroxynaphthalene, which decomposes via ring cleavage and various intermediate stages to o-hydroxybenzaldehyde and salicylic acid, to aliphatic ketonic acids and finally to carbon dioxide [108, 124, 125] (Fig. 3).

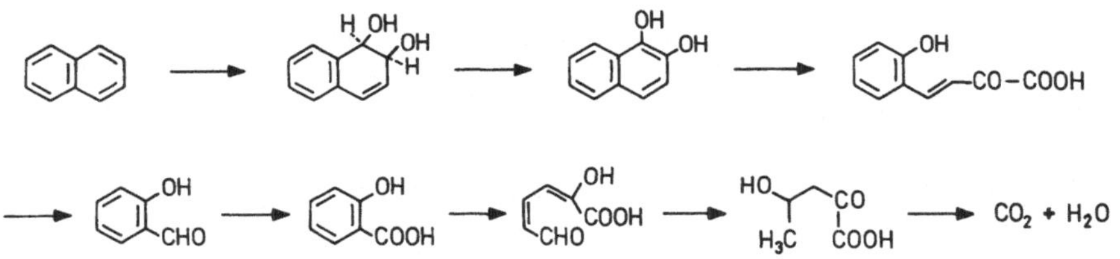

Fig. 3. Naphthalene metabolism by *Pseudomonas putida*

Similar decomposition mechanisms have been found with other Pseudomonas strains (including *P. multivorans*) and with monomethyl and dimethyl naphthalenes, for example to methyl salicylic acids and to dimethyl salicylic acids (if both methyl groups were present in the same nucleus of the naphthalene), or to methyl naphthalene carboxylic acids [51]. On the other hand, *Cunninghamella elegans* metabolizes naphthalene to trans-1,2-dihydrodiol, from which 1-naphthol, 1,4-naphthoquinone and 4-hydroxy-1-tetralone are formed as intermediates for further decomposition [126, 127].

Microbial Degradation of Volatile Aromatics in a Natural or a Seminatural Environment (in Water or in Waste Water Sludge)

Of 1 ml of gas oil about 2 mg dissolves in 1 l of ground water [128]. This extract contains about 15% 1,2,4-trimethylbenzene, about 9% m- and p-xylenes, about 7% 1,2,3-trimethylbenzene, about 7% o-xylene, about 6% naphthalene, about 6% o-ethylmethylbenzene and about 50% 40 other identified benzene and naphthalene homologues [128]. After an incubation time of 1 day (25 °C) to 5 days (10 °C) the ubiquitous microflora in ground water start the decomposition of the hydrocarbons [128]. It takes an additional 5 days for the building up of populations of bacterial species capable of totally eliminating them, provided sufficient oxygen is available [128]. Individual components are degraded at different rates [128]. For instance 1,2,4-trimethylbenzene and m- and p-xylene are degraded much faster than 1,2,3- and 1,3,5-trimethylbenzenes and o-xylene [128]. Naphthalene and methylnaphthalenes are intermediate [128]. Twelve strains, from a total of 30 isolates, affect the composition of the aqueous hydrocarbon solutions, but it seems that there are only 4 different degradation routes [128] and mainly substituted benzylalcohols were identified [128]. Also in coastal seawater – for instance at Cape Cod, Massachusetts – it was possible to identify about 50 volatile hydrocarbons including toluene and naphthalene and their homologues with different biodegradability and other different behaviour in coastal marine processes [129]. In plants for biological treatment of waste water it is possible to differenciate at least two processes when eliminating hydrocarbons [130, 131]. For instance dichlorobenzenes, trichlorobenzenes and trimethylbenzenes are absorbed very rapidly within minutes on the activated sludge until saturation [131]. The partition coefficients for nonpolar compounds between the activated sludge and water correlate with the corresponding octanol/water partition coefficients [131]. Volatile compounds – such as benzene, toluene, trimethylbenzene and dichlorobenzene – are relatively easily transferred from the water phase to the atmosphere during aeration [131]. Compared to these processes microbiological degradation of nonpolar hydrocarbons in the mixture of bacteria, fungi and protozoae of activated sludge is slow [131]. It is easier possible for polymethylbenzenes, with oxygen being the limiting factor [131]. Additionally the microbiological degradation is delayed if other compounds – such as sugars – with higher biogredability compete [131].

Concentrations in the Environment. Overall Environmental Fate

The occurrence of volatile aromatics in the atmosphere and surface water listed according to sources and form of pollution has been dealt with in detail in the first Section. This Section therefore deals mainly with immissions in the atmosphere, in waterways and in the soil (groundwater), which vary greatly depending on location (cities, especially narrow streets), industrial areas, work-rooms, gasoline stations, rural areas, natural waters, effluents, or locations which have been the scene of accidents, etc.). Since data have only been available since the early 1970s, it is difficult to state whether and where environmental pollution by volatile aromatics is decreasing, is more or less constant, or is increasing. Unfortunately, nothing is known to what extent volatile aromatics are present in the gas phase in air and as true solutions in water and to what extent they are adsorbed to aerosols or to suspended particulate matter. There is also little known about total immission concentrations of carcinogens (such as benzene, polycyclic aromatics, specific compounds in aerosols and suspended particles etc.), which may have combined or synergistic effects. People are exposed in varying degree to benzene, for example, from chemical manufacturing processes, coke ovens, various combustion gases from fires, petroleum refineries, solvent operations, the storage and distribution of gasoline, urban automobile emissions, urban gasoline service stations, and using self-service gasoline [132, 133], but at the same time to other effective pollutants. For the population of the USA, it has been estimated that the total annual average benzene exposure from the air is more than 32 $\mu g/m^3$ for 80,000 inhabitants, 12.8–32 $\mu g/m^3$

for 200,000 inhabitants, 3.2–12.8 µg/m^3 for 48 million inhabitants, and 0.3–3.2 µg/m^3 for 110 million inhabitants [132]. To this must be added the uptake of benzene from drinking water, which in North American cities contains an average of some 0.2 ppb benzene, but in isolated cases up to 100 ppb and more [132]. Of food in the USA eggs contain about 2,100 µg/kg, Jamaica rum some 120 µg/kg, irradiated meat about 19 µg/kg, and boiled or roasted meat about 2 µg/kg benzene [132].

Air Immissions of Volatile Aromatics

In urban atmospheres, the composition of aromatic pollutants is similar to that of gasoline [12, 13, 133, 134–136], which would suggest this as the source. Other compositions are found in the vicinity of industrial centers or at places of fire outbreaks [133]. Recently relatively high concentrations of benzene were recognized in a rural populations (Ahlen, Nordrhein-Westfalen) near to a carbonization plant [133a]. The air in remote rural areas contains benzene concentrations of 0.001–0.003 mg/m^3 [132, 134a]. In urban areas with a heavy flow of traffic air contains, in addition to some 0.6 mg aliphatic compounds/m^3, about 0.15–0.2 mg aromatics/m^3 [10, 12], including about 0.04 mg benzene/m^3, about 0.08 mg/m^3 toluene, about 0.04 mg xylenes/m^3, and about 0.04 mg higher benzenes/m^3. In Germany 0.001 mg benzene/m^3, 0.002 mg/m^3 toluene/m^3, and 0.001 mg xylene/m^3 were found on the Feldberg, whereas the concentrations in the centre of Frankfurt am Main at 2 p.m. were 0.14 mg benzene/m^3, 0.21 mg toluene/m^3, and 0.15 mg xylene/mg^3 [134a]. The concentrations in the west-end of Frankfurt were about ten times as high as on the Feldberg, but only a tenth of the immissions in the centre of the town [134a]. In Saarbrücken the concentrations in the interior of cars were measured [134a]. Before driving one found 0.2 mg benzene/m^3 and 0.5 mg toluene/m^3, after 30 min 0.6 mg benzene/m^3 and 1.0 mg toluene/m^3 in the interior [134a]. The benzene portion is high in relation to the gasoline composition, because on the one hand some of the volatile aromatics are cracked to benzene in the gasoline engine, and on the other benzene is more stable in the environment than are its homologues. A number of measurements are compiled in Table 2 (p. 20). However, the data can only be compared to a certain extent, since locations and sampling times, and the sampling and analytical methods applied differ widely. The table can therefore only provide a pointer to the relevant degree of pollution (Table 2). Besides benzene, toluene, xylenes, higher alkyl benzenes one finds also benzaldehyde, methylbenzoate, acetophenone, naphthalene, vinylbenzylalcohol, ethyl benzaldehyde, phenol, dimethylphenol etc. [134].

The extremely high values in Sheffield and London are due to the fact that measurements were taken in narrow streets on a level with automobile exhausts. In the case of the industrial sites, air samples were taken at a distance of approximately 100 m from the factory. Considering the vertical distribution, the maximum toluene and xylene concentrations are found at a height of 140 m, but here temperature inversion situations also play a role [150]. Environmental research into pollution by volatile aromatics as a result of vehicular traffic is being conducted in numerous locations (see, for example [13, 142, 151–155]). Other groups are engaged in examining the transport of pollutants [66] into unpolluted areas and any

Table 2. Immission values for volatile aromatics in mg/m^3

City	Benzene	Toluene	Xylenes	Higher alkyl benzenes
Los Angeles 1965/67 average values [13, 137]	0.05–0.10	0.11–0.17	0.10–0.12	0.10
Los Angeles 1965/66 maximum values [13, 137, 138]	0.18	0.42	0.38	0.36
Los Angeles 1967/71 average values [13, 139]		0.10–0.12	0.08–0.12	0.10
Azusa, Calif. 1967/71 average values [12, 13, 139]		0.04–0.06	0.03–0.06	0.05
Zurich, average values of 8 locations, 1972 [12, 135]	0.04	0.06	0.02	
Zurich, max. values Milchbuck 1972 [12, 135, 136]	0.08	0.15	0.05	
New York, New Jersey [140]		0.15	0.15	
Sheffield, London [12, 141]	0.4	0.9	0.7	
Sendai, Shiogama, Tagajo 1974 [142]	0.06–0.08	0.02	0.06–0.39	0.02–0.07
Dallas, Los Angeles, St. Louis, Chicago 1976 [132,143]	0.003–0.014			
Columbus, Ohio 1978 [132, 144]	0.004–0.030			
Newbury Park, Calif. 1978 [145]	0.005–0.020	0.01–0.04		
Newark Industrial Site N.J. [133]	300	and traces of chlorobenzene		
Bound Brooke Industrial Site, N.J. [133]	9	and 20 mg chlorobenzene/m^3		
Torrance, Industrial Site, Calif. [133]	13.6	and 20.7 mg chlorobenzene/m^3		
Other industrial sites N.Y., N.J., California [133]	2–3	and traces of chlorobenzene		
The Hague 1975 [146, 147]	0.03	0.07	0.07	0.06
Paris 1974 [146, 148]			0.003–0.01	0.007–0.025
Houston 1974 [146, 149]	0.004–0.05	0.001–0.04	0.04–0.07	
Pretoria, Johannesbourg, Durban 1975/76 [146]	0.007–0.02	0.03–0.05	0.02–0.03	0.03–0.04
Frankfurt 1980 [134a]	0.005–0.14	0.005–0.21	0.003–0.15	
Berlin 1978 [134b]	0.002–0.015			

fluctuations in concentration [20, 156]. In addition to the chemical and petroleum industries, other significant sources of industrial pollution are printing operations, the shoe industry, the furniture industry and certain plastics processing operations (see, for example [41, 157–161]). The smoke from forest fires contains substantial amounts of volatile aromatics (contents of 25–320 mg benzene/m^3, 97–891 mg toluene/m^3, and 116–684 mg xylenes/m^3 [13, 133]). In all cases odour problems may be an undisered side effect, which, however, also warns of immissions.

Pollution of Surface Waters

Surface waters contain in addition to other organic compounds relatively small quantities of volatile aromatics, and these are important mainly when they change the taste and smell of the water [13]. Even though the presence of benzene, toluene, xylenes and higher alkylated benzenes can be proved by analytical methods [13, 67, 162, 163], very little information is available on quantities (see Table 3). To date, 65 different non-halogenated benzenoid hydrocarbons have been identified in surface and drinking waters [162]. More relevant are the chlorinated derivatives of these compounds, which, on the one hand, are introduced directly into surface waters or are formed during chlorination of water from the hydrocarbons (with the exception of benzene itself), on the other [67, 162–164].

Of the non-halogenated aromatics, benzene may contribute about 1% to the mixture of 1 mg/l carcinogens occurring in American drinking water [165, 166], which contains an average of about 0.2 µg/l benzene, but in isolated cases up to 0.1 mg/l [132, 166]. Water tests were carried out particularly in Miami (Florida) and Jacksonville (North Carolina), where changes in taste had been established due in part to benzene homologues [163]. In 1976, samples of the River Rhine at Basle yielded an average of 0.2 µg benzene and 0.8 µg/l toluene, at Cologne about the same, and at Duisburg 0.8 µg/l benzene and 1.9 µg/l toluene [167]. The river Glatt, a tributary of the Rhine, was also found to contain benzene, ethyl benzene, xylenes, and higher alkylated benzenes [163]. According to a WHO study, maximum surface-water pollution rates for naphthalene were found to be 3 µg/l and for styrene 1 µg/l [168].

As mentioned, pollution by chlorinated benzenes is more significant. Of these, 1,4-dichlorobenzene is generally the component which occurs most frequently [163]. However, chlorinated toluenes and xylenes are also found, particularly since these hydrocarbons are chlorinated relatively easily during disinfetion [163]. To date, 183 different halogenated aromatics have been identified in waters [164]. The drinking water in Miami contains about 1 µg chlorobenzene/l, about 1.5 µg chlorotoluene/l, and about 2 µg dichlorobenzenes/l [163]. The Rhine contains about 14 µg/l low-volatile chloro-organic compounds (di- and trichlorobenzenes), of which about 4 µg/l pass into the bank-filtered water, and from there about 1 µg/l into the drinking water [169]. Tap water in the Netherlands can contain 0.3–10 µg/l dichlorobenzenes, the taste of which is easily recognizable [170]. In addition to chlorinated aliphatic compounds and chlorinated alkylated benzenes, chlorobenzene and especially 1,4-dichlorobenzene have been identified in effluents, the river Glatt and in Lake Zurich [163, 171, 172]. An interesting point is the depth distribution in a lake; taking Lake Zurich as an example, maximum concentrations in summer of 0.04 µg/l 1,4-dichlorobenzene are obtained at a depth of 10 m, presumably because elimination from the upper layers is effected by transport into the atmosphere [171, 173]. Calculations have revealed that the average residence time of the water in Lake Zurich is 1.2 years, and of the 1,4-dichlorobenzene about 5 months, so that drastic increases in concentrations are unlikely [171]. According to a WHO study, maximum surface-water pollution rates throughout the world were found to be 60 µg/l for o-dichlorobenzene, 200 µg/l for p-dichlorobenzene, and 5 µg/l for trichlorobenzene [168]. Drinking water contains up to 1 µg/l o-dichlorobenzene, 1 µg/l p-dichlorobenzene and 0.1 µg/l trichlorobenzene [168]. Some chlorinated benzenes – such as chlorobenzene – are relatively easily decomposed in part to polar substances, such as o- and p-chlorophenols [71]. In summary the concentrations of these chlorinated mononuclear aromatics are in general very low, but they may

Table 3. Concentrations of volatile aromatics in American surface, tap, and drinking waters[a]

	Surface water from Florida Bay: Sediment + midtreatment + industrial waste water + lagoon effluent [175]	Surface water + final municipal discharge water + tap water in Illinois (Chicago, Ohoi River...) and Indiana [176]	Surface water from Wisconsin River, Wabash River, Kanawha River, Lake Erie, Ocoee River, Chattanooga, Creek [176, 177]	Surface water from Delawere River, Lehigh River, Shuylkill River, Hudson River [176]	Surface water from Mississippi River [176, 178–180]	Drinking water and tap water in Louisiana (New Orleans) [178, 179, 181][b]
Benzene		1– 7 µg/l	0.1– 2 ug/l	Ca. 1 µg/l	Confirmed or probable	Confirmed or probable
Toluene		1– 5 µg/l	1– 2 µg/l	1–2 µg/l	Confirmed, up to 3 µg/l	Confirmed, up to 12 µg/l
Xylenes	2– 8 µg/l	1– 4 µg/l	1– 2 µg/l		Probable	3–8 µg/l
Ethylbenzene			0– 1 µg/l			1–2 µg/l
Styrene	3–30 µg/l				Probable	Probable
Isopropylbenzene	Probable				Probable	
Biphenyl	Probable		0– 2 µg/l		Probable	
Chlorobenzene		1– 4 µg/l		Ca. 1 µg/l		Probable
Dichlorobenzene		1–16 µg/l	0– 3 µg/l	1–10 µg/l	8–15 µg/l	Less than 3 µg/l
Trichlorobenzene	Probable					
Naphthalene	Confirmed, up to 50 µg/l		0–10 µg/l		Probable	Confirmed
Methylnaphthalene	2–30 µg/l		Probable, 1 µg/l		Probable	Confirmed
Dimethylnaphthalenes	Probable, up to 15 µg/l		Probable, up to 3 µg/l		Probable	Probable

[a] From computer outputs of "Water DROP" (Distribution Register of Organic Pollutants in Water). This is a computer information system of the US Environmental Protection Agency and the US National Institutes of Health, which contained 1979 about 4,800 occurences of about 1,500 chemicals.

[b] In tapwater in Prague (CSR) around 0,1 µg/l benzene, less than 0.1 µg/l toluene, and 0.3 µg/l xylenes were found [182]

change the odour and/or the taste of the waters [168]. 1,3- and 1,4-chloronaphthol are formed by the decomposition of 1-chloronaphthalene [174].

Pollution of Soils and Groundwater

Contaminants may reach soils either by precipitation-transfer of atmospheric pollutants or from man-made wastes which are introduced into soils either deliberately or as a result of certain activities – for example, traffic systems or waste tips – but also by seepage of surface waters. Various phases can be detected in individual layers and changes take place in the soil erosion zone [183]. An important factor is the interaction with microorganisms; pollutants can inhibit their growth on the one hand, but many microorganisms contribute to the elimination of pollutants through biodegradation [183, 184]. The question arises as to the loading that can be imposed on soils with respect to processes such as dispersion, adsorption and decomposition [185]. Organic pollutants remain far longer in soil than in water, e.g. chlorobenzene 40 times and dichlorobenzene 200 times longer [185]. Investigations have been carried out to establish the extent to which the filtering soil may be polluted when the groundwater level is lowered and the groundwater replaced by seepage of effluent and running water [186]. For example, the Rhine at Duisburg shows a TOC value (Total Organic Carbon) of 5 mg/l, while the bank-filtered water has only half that amount [185], which means accumulation in the soils. Many investigations have been carried out on the effect of organic pollutants on soil fertility and on soil microorganisms [13], but none specifically concerning the effect of volatile aromatics.

The protection of groundwater from oil pollution has been the subject of various publications [187], but although oil contains aromatics none of these papers deals specifically with aromatics. In Britain, studies have been carried out on the pollution of waters by petroleum fractions (including those containing aromatics) which seep into roads or are washed out [188]. Finally, extensive investigations have been carried out on a case of benzene seepage which occurred around 1967 in Ludwigshafen, but which was not discovered until 1974 [189]. A comparison of the 1974, 1976, and 1978 contamination data measured in the groundwater and soil air revealed that concentration levels had fallen only minimally [189] and that relatively minor geographical shifts had occurred in the pollution zones [189].

Pollution of the Sea

The approximate annual quantities of benzene, toluene, xylenes, and naphthalenes introduced into the sea have been stated in the first Section. Because of their lower volatilities, naphthalene and its homologues in particular tend to concentrate in the biota. In 1974, the USA alone recorded 13,966 oil accidents, of which 545 cases alone involved a total of 4 million l gasoline, and 1,833 accidents involved a total of 4.2 million l diesel oil [190].

Natural coastal waters contain about 1.5 ppb hydrocarbons. Higher pollution rates containing about 25%–36% aromatics are found for example in the Pacific along the Vancouver/San Francisco/Panama Canal shipping route [191]. There are no significant differences between the concentration levels on the surface and at a depth of 10 m [191]. Water dissolves about 9–19 mg/l fuel oil [101]. Of the dissolved portion, about 10%–20%, or 1.5–2.5 mg/l, are methyl naphthalenes and naphthalene [101]. In contrast with other pollutants, these practically do not decompose in a period of four days [101]. About 2,000 t oil are introduced annually into the North Sea, of which 20%–25% have a boiling point below 160 °C [192]. A pollution rate of less than 1 mg/l already constitutes a hazard to larvae and plankton in particular [192]. Measurements in 1974/75 at 47 locations in the Mediterranean showed the presence of dissolved aromatics in amounts from 2–1,055 ng/l (mean value: 167 ng/l) [193]. In Japan, 1,1-phenyl-xylyl-ethane and diisopropyl naphthalenes as PCB substitutes [45] were found. Although these compounds accumulate less and decompose more rapidly, 0.019–0.16 ppm diisopropyl naphthalenes (but no 1,1-phenyl-xylyl-ethane) have been identified in marine sludge near Osaka [45]. Fish accumulate diisopropyl naphthalenes less than naphthalene or dimethyl naphthalenes, but accumulation varies with different isomers [45].

Uptake, Bioaccumulation, Excretion

Aromatic hydrocarbons are biochemically active and more irritating to mucous membranes than are aliphatic and alicyclic hydrocarbons in equivalent concentrations [13]. Toluene and its derivatives penetrate through the cell membrane by simple diffusion [194]. Studies of man and land mammals have focussed on the uptake and excretion of benzene, toluene, xylenes and styrene [10, 12, 13, 132, 136, 195–198], which are mainly inhaled (and exhaled again) or involve skin contact. The skin can resorb 0.4 mg benzene/cm^2/h [198]. In the case of algae, molluscs, fishes, and other aquatic organisms, the uptake of less volatile and thus accumulating aromatics – such as naphthalenes – is of greater significance. Concerning excretion mechanisms, reference is made to the next section "Metabolism".

Uptake and Excretion of Benzene, Toluene, Xylenes and Styrene by Man and Land Mammals

About half of the inhaled benzene is exhaled again in less than 20 minutes [11, 12, 136, 196, 197, 199]. Given longer exposure times, the resorption rate is initially higher and then decreases [197]. Depending on the individual, resorption can be as high as 80% [195]. The resorbed benzene is deposited by the blood in the central nervous system and the bone marrow (18 times the blood concentration), in the fatty tissue of the peritoneum (10 times the blood concentration), in the heart (5 times the blood concentration) and in the brain (2.5 times the blood concentration [12, 136, 197]. The distribution coefficient between air and blood is about 1:7 [197]. The benzene retention time in the system is quite long [197]. Approx. 80% have been oxidized and removed after three days, but even after 5–7 days traces of benzene can still be detected – for example, in the bone marrow, testicles, and the skeletal muscles [197]. The resorbed benzene is excreted almost entirely in the form of phenol in the urine [195, 197], a fact which is used in occupational medicine for health surveillance.

Many scientific investigators (e.g. [200–203]) have studied the kinetics of resorption, of elimination and of metabolisation of benzene and toluene. A study of employees at petrol stations showed that with an average load of the working air of between 0.065 and 0.490 ppm benzene within eight hours (with a maximal 30 min value of 4.462 ppm; the german technical threshold limit for the average 8 h day is fixed at 8 ppm) it was possible to find between 6 and 37 mg of benzene and its metabolites in 1 l of urine, which is in the order of magnitude of persons not exposed to benzene [204]. There was no correlation between benzene concentrations in gasoline, in the air at the petrol stations and the phenol concentrations in the urine of the employees [204]. In another study with volunteers the elimination of benzene, which was inhaled up to five days in concentrations of 2–10 ppm was differentiated between exhalation and excretion in the form of phenol [205]. Females seem to eliminate benzene slower than males [206, 207]. Also rats and men with raised fat contents are eliminating benzene slower [206, 207]. In another study with female workers in the rubber industry working with gasoline, it could be shown that hydrocarbon fractions (which were not differentiated) accumulate mainly in the embryo and in the mother's milk [208].

In the case of toluene inhalation about 20%–50% is exhaled and about 50%–80% resorbed [195, 197]. An increase in the air concentration to 200 ppm (0.75 mg/l) corresponds to an increase in the blood concentration of about 0.35 mg/100 ccm [197]. Toluene is also mainly absorbed by the lipid-containing organs such as the suprarenal gland, bone marrow, brain, liver, blood, etc. [197]. Most of the toluene is converted to benzoic acid, which after conjugation with glycine is excreted with the urine as hippuric acid [12, 136, 195, 197]. Since the body also has a normal, varying hippuric acid excretion rate, it is difficult to weigh the contribution by toluene [195]. Toluene is excreted more rapidly than benzene, total elimination being attained after about 14 h [197]. The biological half-life of toluene is about 7½ h, and is hardly changed by other solvents [209].

Similarly to toluene, xylenes are taken up by the lungs, through the skin or the mouth and pass into the lipid-rich tissues [197]. They are oxidized to toluic acids and xylenols, which are excreted with the urine in various forms – possibly as conjugates – with some of the fractions requiring somewhat longer than in the case of toluene [197].

When styrene is inhaled, about two-thirds are resorbed; it can also be absorbed through the skin [197]. It accumulates in all organs, but principally in the liver, kidneys and blood and is also found in faeces [197]. It is oxidized at the double bond and decomposed to mandelic acid and phenyl glyoxylic acid which are easily identifiable in the urine [195]. Benzoic acid has also been detected in rabbits [197]. As with the xylenes, excretion takes about three days [197]. If guinea-pigs are exposed to only about 1 ppm styrene (0.005 mg/l), however, no mandelic acid is detected in the urine – only at higher concentrations [197].

Mathematical models concerning the uptake, transport and effect of nonionic poisons (benzene in particular) have been developed and compared with the results of experiments [136]. These revealed that the concentrations are purely additive [136].

Uptake, Excretion and Toxicity of Volatile Aromatics in Aquatic Organisms*)

Concentrations of less than 1 mg/l oil in water are sufficient to kill the larvae of all animal species, phyto- and zooplankton [192]. 1–10 mg/l constitute the danger level for crustaceans and worms, and 1–50 mg/l for fishes, oysters, and mussels [193]. Snails are more resistant and do not die until the degree of pollution has reached 10–100 mg/l [193]. In one of the models already mentioned, a comparison was made of the decomposition of benzoic acid, chlorobenzene, nitrobenzene and hexachlorobenzene in algae, daphnia, mosquitoes, snails and fishes [71]. Benzene, toluene and xylenes are evidently quite toxic for phytoplankton, but experimental proof is difficult since these aromatics are so volatile that they easily escape from the testing installation [210]. Benzene is about ten times more toxic for daphnia (LC$_{50}$ 5 ppm) than for fishes (LC$_{50}$ 50 ppm) and algae, with mercuric chloride and/ or detergents synergistically increasing the risk to daphnia to TC$_{50}$ 0.2–0.5 ppm [211]. Benzene concentrations of 10–20 ppm hinder the growth of "Skeletonema sp." [211]. The uptake of chlorinated naphthalenes by marine unicellular algae has also been studied [212]. For p-dichlorobenzene see [292].

The relative toxicity of an oil to marine organisms is, in most cases, directly correlated to its content of aromatic hydrocarbons [48]. The acute toxicity increases with increasing molecular size from benzene to phenanthrene with the alkyl analogues being more toxic than the parent compounds [48]. Estuarine and benthic species are often more tolerant than marine species, and larvae and juveniles of many species are more sensitive than adults [48]. For grass shrimps, for example, LC$_{50}$ (96 h) are 27 ppm benzene, 9.5 ppm toluene, 7.4 ppm xylenes, 2.4 ppm naphthalene, 1.1 ppm methyl naphthalenes, and 0.7 ppm dimethyl naphthalenes [48]. The bioaccumulation factor in the case of the "Rangia cuneata" mussel is 6.1 for naphthalene. 66% of the naphthalene are eliminated within 24 h [48]. Exposure rates as low as 0.5 ppm naphthalenes roughly double the respiratory rate of crustaceans and roughly halve the growth of worms and crustaceans [48]. The "Macoma inquinata" mussel picks up naphthalene not directly from ingesting the sand in oil-soaked sediments but by absorbing the organic portion which has passed into aqueous solution in the sediments [46]. In summary naphthalene and its homologues are relatively toxic to oysters and mussels and are – compared to other hydrocarbons – excreted only very slowly [48].

Several studies have been made on the uptake of aromatic compounds by fish: following exposure over several weeks, the tissues of rainbow trout contain naphthalene in concentrations of 40–300 times and in the gall of 13,000–23,500 times that of the water [213]. Salmonoids take up aromatics – anthracene more than naphthalene, and naphthalene more than benzene – via the gills or through the skin, and these pollutants are deposited primarily in the gall bladder, the liver and the brain, where they are metabolized [47]. Concentrations of as little as 0.5 ppm naphthalenes in seawater lead to pronounced accumulations in the flesh within a

---

*) A very valuable table of acute toxicity of benzene, toluene, xylenes, ethylbenzene, trimethylbenzenes, naphthalene and methylated naphthalenes to freshwater and marine animals (bay shrimp, pink salmon try and striped bass are especially sensitive) is given by. J. M. Neff "Polycyclic Aromatic Hydrocarbons in the Aquatic Environment", table 82. Applied Science Publ. 1979. Naphthalenes are about 10 times as toxic as benzenes.

matter of 2–6 weeks, but if followed by a period in clean sea water are completely eliminated [47]. Elimination is effected mainly through the gall/liver/faeces tract [49]. Naphthalenes are also washed out through the fish skin [47]. However, metabolites are retained for longer periods in the liver and skin [49]. Flat fish – starry flounder and particularly rock sole – accumulate naphthalene metabolites even more than do salmonoids [49]. Goldfish accumulate diisopropyl naphthalenes less than naphthalene and dimethyl naphthalenes, with different diisopropyl naphthalene isomers having different retentions [45].

When "Cirolana borealis" is exposed to 1 ppm toluene, practically no effects are noted [214]. At 5.7 ppm, the swimming activity increases at first and the contents of the stomach are expelled through the mouth [214]. After 400 h, the fishes are totally inactive and lie on their backs or sides [214]. Various methods to determine the behaviour of "Cirolana borealis" under conditions of stress when exposed to toluene have been compared [214].

Metabolism

The decomposition of benzene and naphthalene and its homologues by microorganisms has already been discussed earlier. The metabolizing mechanisms of naphthalenes in fish have been well studied [47, 49]. Decomposition products of chlorobenzene in daphnia, mosquitos, snails and fishes are the polar compounds chlorophenol and chloro-o-dihydroxybenzene amongst other compounds, those of nitrobenzene aniline, acetanilide, aminophenols and nitrophenols and those of hexachlorobenzene pentachlorophenol and unknown compounds [71]. Bromobenzene is deactivated to the toxic bromophenol [217]. In the case of man and land mammals, studies have concentrated on the metabolism of benzene, toluene, xylenes and styrene, which are also significant in occupational medicine [12, 13, 136, 195, 196, 215–217]. A comparison of the metabolism of benzene into phenol in various animal species with the aid of microsomal preparations of the lungs or liver yielded vast differences. However, it is possible for benzene, in part, to inhibit or prevent its own metabolism [218].

The Metabolism of Benzene in Man and Mammals

Aromatic hydrocarbons are metabolized both in vitro and in vivo to arene oxides, which isomerize to phenols, are enzymatically hydrated to dihydrodiols, and are conjugated with glutathione [12, 13, 196]. The dihydrodiols are further metabolized to catechols by dehydrogenation [12, 13, 196]. Figure 4 shows a simplified diagram of the correlations applicable to all volatile aromatics [12], while Fig. 5 gives a more detailed diagram of the metabolic pathway of benzene in the human liver [196].

The initial step of the oxidation of benzene to the epoxide has been widely studied (see [196, 217, 219, 220]): It is the reaction mediated by a mixed-function oxidase [196, 219] in the NADPH/NADP-system of the microsomes [220]. Cytochrome P-450, NADPH-reductase and phosphatide are important factors

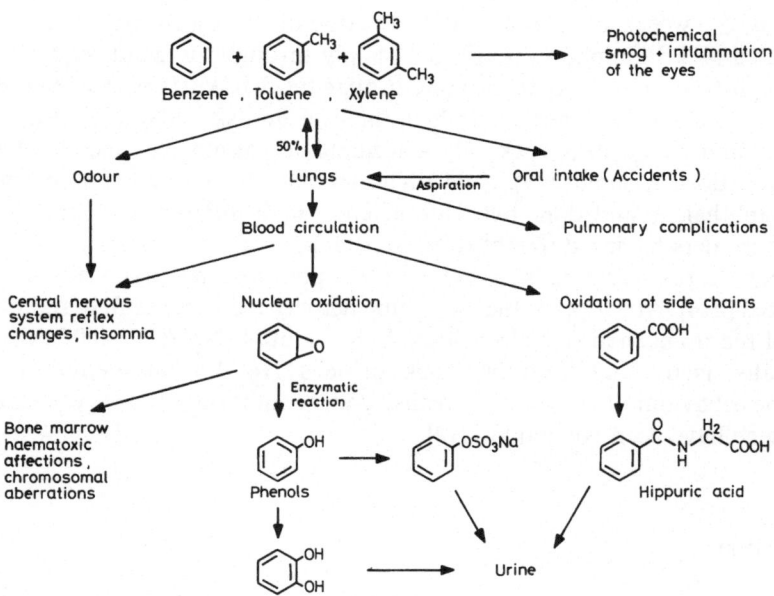

Fig. 4. Scheme of fate and effects of benzene, toluene and xylene in the body [12]

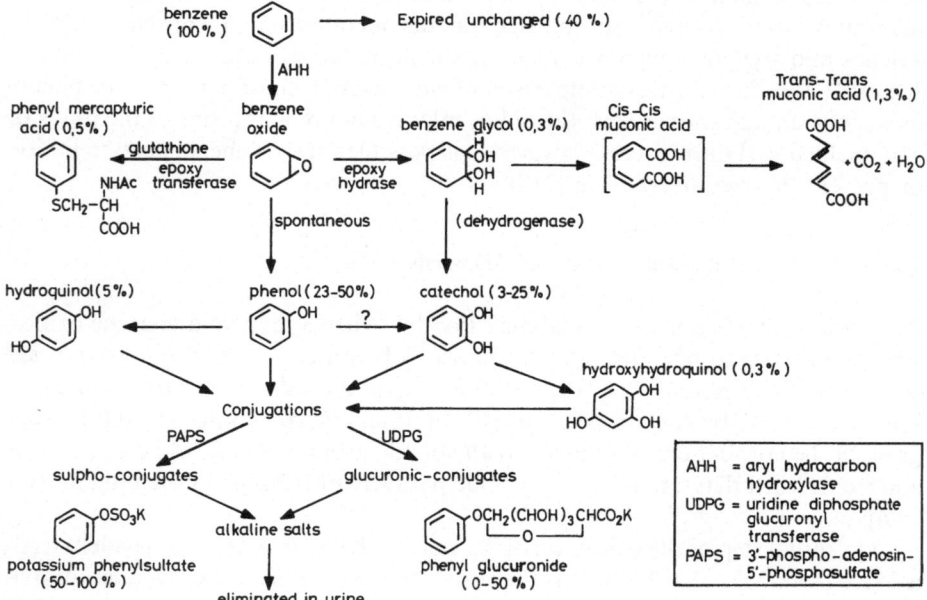

Fig. 5. Metabolism of benzene in man [196]

[217]. Phenobarbital and 3-methylcholanthrene can increase the rate of benzene metabolism [196, 219, 219a]. Toluene may retard the metabolism of benzene in liver microsomes of mice [220]. Since benzene also reduces the incorporation of iron, this reaction could also be a measure for benzene uptake [220]. The detoxification process of the highly reactive [196] carcinogenic [12, 215] and liver injuring [12, 215] arene oxide leads to dihydrophenols and to dehydrocyclohexanon [217] as intermediates. Recently, it has been shown that reactive aromatic epoxide compounds – which can form covalent bonds with DNA, RNA and/or proteins – react non-enzymatically to phenols or enzymatically together with epoxidase hydratase or GSH-S epoxide transferase to dihydroxy compounds or to monohydroxy glutathione conjugates [216]. Additional enzyme activation is possible with progressive development in the animal phylogenetic development [217]. It seems also that there are reactive metabolites, of which one leeds to reparable damages and the other to irreparable toxic covalent binding [220]. It is interesting to note that benzene toxification of rabbits accelerates also the synthesis of nucleic acids in the bone marrow [12, 221].

The Metabolism of Toluene and Xylenes in Man and Mammals

Reference has already been made to the decomposition reactions of toluene (see Fig. 6) and xylenes (see Fig. 7) [12, 136, 195, 197]. Such alkyl benzenes are metabolized in vivo both to phenols and to products resulting from side-chain oxidation [13]. Thus, toluene is metabolized primarily to benzoic acid via the intermediate formation of first benzyl alcohol and then benzaldehyde [13]. Both o-cresol and p-cresol are also detected as minor in-vivo metabolites [13]. Xylenes are metabolized in vivo primarily by oxidation of a methyl group to yield toluic acids [13]. It has been suggested, that a small amount of toluene or xylenes may also be hydroxylated in the benzene nucleus, when administrated to male albino rats [12, 222].

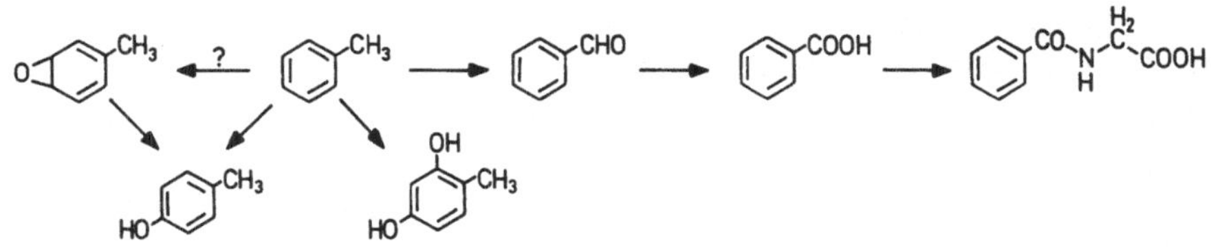

Fig. 6. Metabolism of toluene in mammals

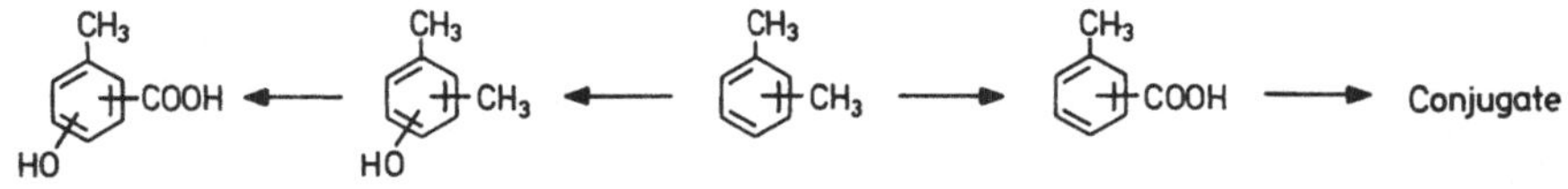

Fig. 7. Metabolism of xylenes in mammals

The Metabolism of Styrene in Man and Mammals

A large number of studies have been carried out on this subject in connection with the production of plastics. As mentioned earlier benzoic acid, mandelic acid, phenylglycol and phenyl glyoxylic acid are formed [195, 197] (Fig. 8).

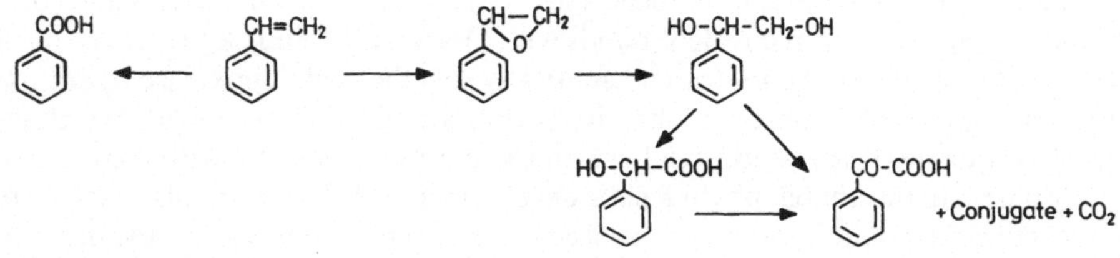

Fig. 8. Metabolism of styrene in mammals

The Metabolism of Naphthalenes

Naphthalene-1,2-oxide is the obligatory intermediate in the microsomal hydroxylation of naphthalene to 1-naphthol, *trans*-1,2-dihydroxy-1,2-dihydronaphthalene and a glutathione conjugate [13, 23] (Fig. 9).

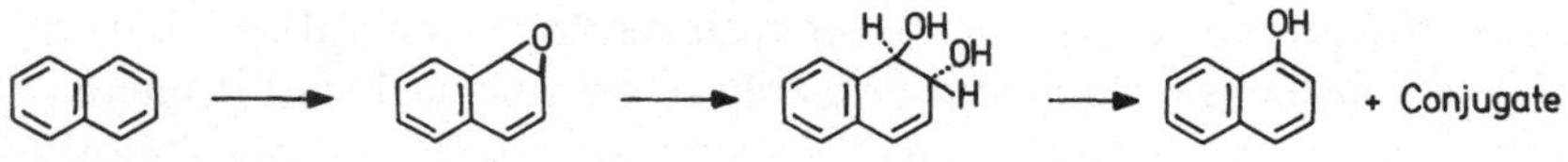

Fig. 9. Metabolism of naphthalene in mammals and fish

Studies of various species of fish have shown that these metabolize naphthalene in the same way as rats and other mammals [47, 49]. Eight different derivatives of naphthalene have been identified [47, 49], two of these being non-conjugates (1-naphthol and 1,2-dihydroxy-1,2-dihydronaphthalene) and four conjugates (1-naphthyl glucuronic acid, 1-naphthyl mercapturic acid, 1-naphthyl sulphate and 1-naphthyl glucoside) [47, 49]. Contrary to the metabolism in certain microorganisms (see Fig. 3), arene oxides are involved (see Fig. 9) in this instance, which results in the formation of trans-diols [47, 49, 108–110]. The relationship between the metabolite concentrations varies enormously in different fish species and in the various organs [47, 49]. The brain of trout contains almost only non-conjugates [47], while the gall of flat fish contains almost only conjugates [49]. Varying quantities of all derivatives are found in the liver, skin and urine [47, 49]. Only two derivatives from the decomposition of 2,6-dimethyl naphthalene (see Fig. 10) have

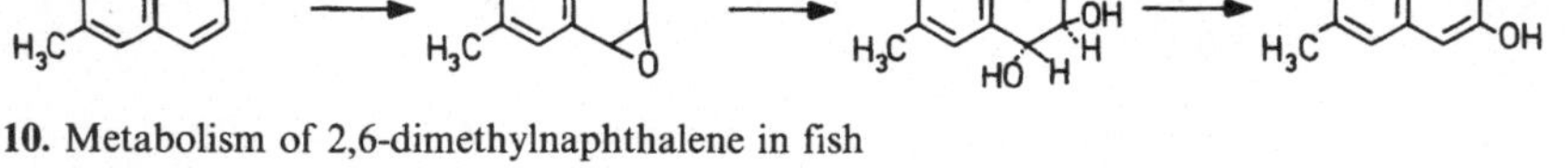

Fig. 10. Metabolism of 2,6-dimethylnaphthalene in fish

been identified to date, i.e. *trans*-3,4-dihydro-3,4-dihydroxy-2,6-dimethyl naphthalene and 3-hydroxy-2,6-dimethyl-naphthalene [47]. It is not yet understood why in this case the β-position is preferred for the hydroxy group.

Biological Effects and Toxicity in Mammals*)

General Remarks, Acute Toxicity of Benzene

Numerous investigations have been carried out on the acute and chronic toxicity of volatile aromatics (for reviews, see: [3, 10, 12, 13, 132, 137, 195–199, 224–228]). Four newest reviews should especially be mentioned:
- Lawrence Fishbein "Potential Industrial Carcinogens and Mutagens, Chap. 22: Aromatic Hydrocarbons (with a subchapter of 10 pages about benzene, and a subchapter of 2 pages about toluene, and 77 literature references)," Studies in Environmental Science 4, Elsevier, 1979
- R. Lauwerys "Industrial Health and Safety, Human Biological Monitoring of Industrial Chemicals 1: Benzene" with 46 pages (including 5 figures and 84 literature references), Commission of the European Communities EUR 6570/1979
- Benzene: "Interpretation of Data and Evaluation of Current Knowledge" (Workshop Vienna, June 1980), Bericht Forum für Wissenschaft, Wirtschaft und Politik e.V., Bonn, Heft Nr. 9 (March 1981)
- DGMK-Research Report 174-6 "Evaluation of Benzene Toxicity in Man and Animals", Deutsche Gesellschaft für Mineralölwissenschaft und Kohlechemie e.V., Hamburg, June 1980

A strongly simplified diagram of toxic effects as a function of dosage is shown in Table 4 [12]. Apparently, no human health effects have been reported for benzene concentrations below 10–25 ppm [12, 13, 196–198, 224, 229], and opinions still differ on the dose-effect relationships. Man and mammals are particularly sensitive with regard to certain effects of benzene on specific enzyme systems and neurophysiological reactions in the brain, and also about odour and taste perception [12, 13, 196]; this has resulted in very low tolerance limits being set in eastern Europe. After administration of 2–3 ml/kg benzene or gasoline liveralkaline phosphatase was increased by over 200%, whereas kidney alkaline phospatase activity was depressed by 50% [229a]. Since it has recently been established that additional peripheral lymphoctes, chromosome breaks, dicentric chromosomes and translocations occur as a result of long-time exposures to as little as 2–3 ppm benzene [196, 230, 231], and since these changes could be precursors of leukaemia [196, 228], western countries are also gradually lowering their threshold limit values (Federal Republic of Germany, MAK-value 1966: 25 ppm, 1970: 10 ppm, 1971: no MAK-value because carcinogenic; USA 1934: 75 ppm, 1948: 35 ppm, 1974: 10 ppm; NIOSH recommendation 1977: 1 ppm; Swiss MAK-value 2 ppm; Sweden 5 ppm [28, 195]. According to a more recent analysis, it is estimated that about 1% of all leukaemia deaths in the USA are due to environmental pollution with benzene [228]. It was, however, pointed out that we do not know of any evidence that leu-

---

*) Effects on microorganisms, plants and fish see earlier section

Table 4. Order of magnitude of the toxic doses by inhalation of volatile aromatics for man and mammals (G. Fodor, 1972, G. Büttner, 1974, E. Merian, 1974, and more recent literature)

	Benzene	Toluene	Xylene	For comparison: Lead
Acute lethal toxication	15–65 g/m³	5–50 g/m³	10–60 g/m³	800 mg/m³ [a]
Poisoning, unconsciousness, mucositis, dyssomnia, disturbances of the central nervous system	0.6–3 g/m³	0.4–4 g/m³	1.3–4 g/m³	0.03–10 mg/m³ [b]
Chronic toxication, metabolic changes, changes in the blood composition, changes in the white blood count, leukocytosis, anaemia, chromosomal changes	at least 0.01–1 g/m³ and more	3–20 g/m³ [c] Many observations without statement of concentration	[c]	
Odour perception thresholds	2.8–5 mg/m³	0.1–2 mg/m³	0.6–0.7 mg/m³	Zero effect level for changes of the ALA-metabolism
Electrical activity of the cerebral cortex; biochemical changes in pregnant mammals	1–2 mg/m³	0.6 mg/m³	0.2 mg/m³	0.004 mg/m³ [d]
Maximum immission concentration values Soviet Union 1960	0.8 mg/m³	0.6 mg/m³	0.2 mg/m³	0.002 mg/m³
West Germany 1966 (VDI/MIK values)	3 mg/m³	20 mg/m³	20 mg/m³	

[a] Source: G. Roush, New Orleans
[b] Symposium Amsterdam 1972 and monograph "Lead" by the National Academy of Sciences, Washington, D.C., 1972
[c] No sure evidence of risks of pure toluene and xylene, because contaminated solvents are normally used
[d] Source: L. Goldberg, Albany, N.Y.

kaemia has developed in a benzene worker whose exposure was discontinued after the appearance of early haematological changes [199]. Attempts have also been made to identify the biological effects of gasoline and diesel oil exhausts (summarized in [24], but see also [232, 233]). However, little is known about what harmful consequences are actually attributable to volatile aromatics. Benzene concentrations in normal outdoor air in Switzerland are, at least 1,000 times less than those which have caused diseases of the blood [10]. Since knowledge about the effects of very small quantities is still most sketchy, it is desirable that the general population be exposed to the lowest concentration possible [10, 234, 235]. Gasoline should not be used as a cleaning agent [10], nor benzene as a solvent [3]. Mixtures containing more than 0.5% benzene are legally treated as benzene in the USA [236]. Legal limits and controls for volatile aromatic hydrocarbons have recently been compared [236 a].

Acute Toxicity of Benzene

Cases of severe, acute benzene toxicity have now become rare [3, 12, 197, 198, 224, 237]. Benzene affects the central nervous system [12, 224, 238], and death occurs as a result of circulatory failure at relatively high exposure rates of 5,000–45,000 ppm benzene [197, 198]. Concentrations of 500–1,500 ppm cause unconsciousness [198, 239], those of 150 ppm a state of euphoria [3]. The euphoric component can, in persons so disposed, lead to benzene addiction [198]. Taken orally and following aspiration, benzene at an LD_{50} of 1–6 g/kg [3, 197, 224] causes nonspecific pneumonia [3]. In the case of acute toxicity, no changes are noted in the blood when examined under the microscope – even if the benzene content is 0.1–2 mg/100 ccm [198]. Acute toxicity gives no indications of the chronic effects [198]. Benzene has practically no warning effect, and its odour is not unpleasant [3]. Eye irritations do not occur until exposure rates of approx. 3,000 ppm benzene [3].

Chronic Effects of Benzene: Cytogenicity, Carcinogenicity, Leukaemia, Pancytopenia, Chromosomal Aberrations

The accumulation of benzene metabolites in bone marrow along with the coincidental covalent linkage of benzene to solid residues of bone marrow is consistent with a phenomenon of toxico- and carcinogenesis shared by many chemicals [196]. All the available documentation strongly suggests a relationship between chronic benzene exposure, chromosome damage and leukaemia. Benzene is a human leukaemogen [196]. There is no convincing evidence that benzene causes neoplasias – including leukaemias – in animals [196]. This could be due to an as yet unknown human cocarcinogen required to evoke the leukaemogenic response [196]. All risk assessments are based on epidemiological studies, and on the assumption that changes in the blood composition constitute precursors of more serious diseases – such as leukaemia [196–199, 224, 228, 240, 241]. The mentioned Vienna Workshop of June 1980 of the Forum für Wissenschaft, Wirtschaft und Politik e.V. came to the conclusion, that a threshold limit of 10 ppm benzene could be retained as a general recommendation, but that further studies are necessary, including contributions from oral intake of benzene in food, water and tabocco smoke.

Haematoxicity, particularly pancytopenia (diminution of all formed elements in the blood), has been observed in both humans and animals following exposure to benzene [12, 13, 196, 198, 224, 242–248]. Benzene-induced pancytopenia and aplastic bone marrow indicate a greater risk to acute myelogenous leukaemia and erythroleukaemia [196]. The effect of increase in red blood cell levels of delta-aminolevulinic acid (a precursor in the haeme biosynthetic pathway) and the effect of decrease in the mean serum complement of the blood occur at benzene occupational exposure levels of 3–15 ppm, and may also have a potential significance [196]. According to a study of the hypofunction of different blood systems, disorders occur at benzene exposure rates of 2–500 ppm [3, 12, 13, 198]. Isolated cases of abnormal blood composition due to minimum exposure levels of 2–3 ppm have been reported from Italy and France [249, 250].

At exposure rates of less than 1–3 ppm, cytogenetic studies indicated a 10-fold increase in chromosome breaks compared with control persons [196, 230, 231]. Studies with 10 ppm to 150 ppm benzene strongly suggest that chromosome breakage and rearrangement can result from chronic exposure [12, 13, 196, 198, 199, 224, 251]. Normally, such damage occurs at exposure rates of 25–200 ppm [13, 196, 198, 224], but at least in one study significant effects were noted at a time-weighted 8 h average dosage of 2–3 ppm over a period of from 1 month to 26 years [196, 230, 231]. No dose-effect relationship has so far been established for benzene-induced chromosome aberrations [196]. In workers chronically exposed to levels in the range of 5–25 ppm benzene, both positive and negative reports involved small numbers of workers [196]. It seems that the formation of abnormal blood composition requires about 5 months at exposure levels of 100 ppm, about 20 months at 50 ppm, about 50 months at 20 ppm, and about 100 months at 10 ppm [189]. Decrease in DNA synthesis, reactions with proteins, breaks and gaps in chromosomes were also studied in cultured human cells and in the bone marrow of rats and rabbits after treatment in vivo [196, 252]. There is evidence that benzene affects the lymphoid system [199]. In rats, 50 ppm and 200 ppm benzene (8 h per day, 5 days a week, 750 h) causes leukopenia and reduces the lymphocyte and polymorph counts and also the myeloctic activity of the bone marrow [196]. Since a favoured mechanism for leukaemia development is somatic mutation, this supports the thesis that benzene is a leukaemogen [196]. In man, pancytopenia and acute myelogenous leukaemia and their variants are clearly related to benzene [196]. However, it must be borne in mind that the incubation period for leukaemia is 10–25 years [199]. When comparing with other forms of leukaemia, it must also be remembered that there are at least four types of leukaemia with different formation mechanisms and with varied risk, depending on age and population group [199, 228]. Since most studies concern middle-aged to elderly males who were occupationally exposed to benzene, no conclusions can be drawn about differences in susceptibility of other populations [196]. Several studies have been conducted into the effective risks for industrial workers and for the population at large [196, 198, 199, 228, 253–256]. Reports about cases of leukosis after 1 ½–15 years apply to benzene exposure levels of 16–660 ppm [198]. Of the more recent epidemiological investigations, four in particular have become important:
– Study by P. F. Infante [228, 257] 1977, which is, however, controversial [199, 255]. In a retrospective study of mortality in a cohort of 748 white male workers

in two Ohio plants manufacturing a natural rubber cast film product, a statistically significant higher rate of leukaemia than in either of two control groups was observed. The persons were probably exposed to about 15 ppm to 100 ppm in the beginning, and later to about 10 ppm to 50 ppm [228] and lower.

- Study by M. Askoy [228, 258] 1974–1977. He compared the types of leukaemia observed in shoe workers, who work with benzene solvents in small unventilated shops in Istanbul, with the types of leukaemia observed in people with no exposure to benzene. The concentrations outside working hours ranged between 15 ppm and 30 ppm, and during working hours reached a maximum of 150 ppm to 210 ppm when adhesives containing benzene were being used [228].
- Study by M. G. Ott [228, 259] 1973–1977. The long-term mortality patterns and associated exposure estimation of a cohort of 594 workers exposed to benzene were reported. The concentrations ranged from less than 2 ppm to greater than 25 ppm [228].
- Study by D. J. Kilian et al. and D. Picciano [196, 230, 231] 1978. Occupational exposure of 52 workers in the Texas Division of Dow Chemical to benzene for periods of from 1 month to 26 years (average exposure time 56.6 months). The time-weighted 8-h average dosage was 2 ppm to 3 ppm benzene [196].

The population risk to ambient benzene exposures has been extrapolated from the first three of these four studies [196].

It has been calculated from a model that the number of leukaemia cases per annum in the general population of the USA due to atmospheric benzene (see [132]) could be about 90 with a 95% confidence interval [228]. This is about 1% of the total leukaemia deaths in the United States, based on 1973 vital statistics [228].

Other Effects of Benzene: Mutagenicity, Embryonic Effects, Synergisms, Effects on the Nervous System, Eye Reflexes, Skin Lesions

In addition to the risk from leukaemia, benzene exposure is also likely to induce mutations [228]. The magnitude of this risk cannot be estimated because of the uncertain quantitative relationship between heritable mutations and chromosome aberrations which have also been observed [199, 228]. Benzene was found to be non-mutagenic in the Ames test for point mutational effects [199, 228]. However, it could be possible that a human metabolic activation enzyme system would cause it to be mutagenic [228]. The genotoxicity of benzene, toluene, xylenes and phenols has been compared [260]. The few reports of reproductive, embryonic and teratogenic effects of benzene are conflicting and inconclusive, and hence are not useful in evaluating the possible risk [199, 228]. High exposure rates of 1,000 ppm to 5,000 ppm can cause cell deformation in animals [224].

Lower molecular aromatics affect the central nervous system, and are likely in small doses to engender unspecific disorders, such as lassitude, irritability, nausea, insomnia, and reduction of power of concentration and individual performance [12, 198, 261, 262]. Several researchers have attempted to objectively correlate such forms of distress with the aid of neurophysiological and other methods [12, 263–265]. For instance, 1.5 mg benzene/m^3 are already subliminal with respect to their effect on the electric activity of the cerebral cortex [12, 264]. These threshold values approximate with the odour threshold values [12]. On the other hand, certain benzene levels in narrow streets are also of this order [12].

As already mentioned, eye irritation due to benzene, unlike other volatile aromatics, only occurs at relatively high concentrations [3, 13, 266]. Research is also being carried out on the pathogenesis of skin lesions due to benzene [3, 267]. Very important are the synergistic effects between volatile aromatics and other pollutants. It is known, for instance, that phenobarbital, chlorprimazine and 3-methylcholanthrene [196, 219, 219a], toluene [220], iron [220], carbonmonoxide [268], cigaret smoke [269], perchloroethylene [270], selenium [271], vibration and noise [272], and sulfur dioxide as function of the climatic conditions [273] influence the uptake, the metabolism and/or the effects of benzene.

Toxicity of Toluene

Toluene levels of 19,000 ppm are lethal after a short exposure period [197]. Concentrations of 800–1,000 ppm cause severe poisoning [197, 239], those of 50–100 ppm produce faintness, nausea, drowsiness, lassitude and headaches [197, 239]. Such concentrations are not present in the environment [13].

According to present knowledge, toluene as opposed to benzene is not haematotoxic and hence is used as a benzene substitute [195]. Quite a few epidemiological observations of changes in the blood composition [12] are probably not significant, because the victims were exposed to solvent mixtures or to different solvents through the years, and crude toluene often contains benzene [12]. Toluene can possibly lead to changes in the blood composition at concentrations of 3–20 g/m^3, and is therefore much less dangerous than benzene [12]. Nevertheless, numerous investigations have been carried out recently on the effect of toluene on blood properties [243–245, 274–279] and possible genetic effects [260]. Toluene has a stronger neurotoxic effect than benzene [12, 276, 280], and can be identified neurophysiologically in concentrations as low as 0.6 mg toluene/m^3 by measurements of the electric activity of the cerebral cortex [12, 264]. Toluene induces also activation of certain hypthalamic and median eminence catecholamine nerve terminal systems of the male rat [280a]. The odour of toluene is less pleasant than that of benzene [12, 197]. Toluene is also more eye irritating than benzene [13].

Toxicity of Xylenes

In acute cases, xylenes seem to be slightly more toxic than benzene and toluene, in that 2,000–5,000 ppm can already be lethal [197], but otherwise they are very similar to toluene. Xylenes are also frequently used as solvents and are now therefore subject to strict controls. Surveys have recently been conducted concerning the toxicity of xylenes [281–284], including studies into the possible incidence of blood changes [277, 285, 286]. However, like toluene, xylenes are apparently not ambient poisons, and at the most they constitute an occupational risk, including that of contact-dermatitis [287]. However, xylenes have a stronger neurotoxic effect than benzene and toluene [12], and can be identified neurophysiologically in concentrations as low as 0.2 mg xylene/m^3 with measurements of the electric activity of the cerebral cortex [12, 264]. In epidiomologicae studies of persons working with xylene as solvent, non specific disorders and deficient sense of touch were found [288]. As irritants to eyes, xylenes are more powerful than benzene but weaker than toluene [13]. Ethanol accelerates, physical activity reduces the negative impacts of xylene [289].

Toxicity of Ethyl Benzene and Styrene

The toxicological properties of these important industrial intermediates have been very well studied [195, 197, 239]. In acute cases, they exhibit similar toxicities to toluene and xylenes. Concentrations of about 1,000 ppm cause severe poisoning, while 100–200 ppm cause disorders. It is suspected that chronic exposure might possibly demage the central nervous system. Styrene appears not to possess any carcinogenic activity [195]. Ethyl benzene is an irritant to eyes on a similar level with toluene, but styrene is a much stronger irritant [13]. Eye irritations due to ethyl benzene begins at concentrations of about 1,000 ppm; however, the eyes become accustomed if exposure continues over a lengthy period [13]. So far, there is no evidence that styrene is an ambient poison. The induction of nuclear styrene mono-oxygenase and epoxide hydrolase in rat liver has been studied [290]. The acute lethal dose of isopropyl benzene for mice is about 2,000 ppm [239].

Toxicity of Chlorobenzenes and Nitrobenzene

Exposure to 2,000–3,000 ppm chlorobenzene is lethal in mammals [239]. In man, chlorobenzene causes severe poisoning at 400 ppm and disorders at 75 ppm to 200 ppm [239].

In man, o-dichlorobenzene causes severe poisoning at concentrations of 300 ppm [239]. 25 ppm to 100 ppm have a pungent odour and produce disorders [239]. In the case of p-dichlorobenzene, concentrations of 160 ppm produce disorders, and 50–80 ppm are irritating to the eyes [239]. At concentrations of about 15 ppm, neither compound produces any discomfort [239].

Nitrobenzene concentrations of 200 ppm cause severe poisoning while 1–40 ppm produce disorders [239].

Odour of Volatile Aromatics

Changes in odour and taste are frequently more irksome than minor health disorders. Industrial odour immissions by toluene, xylenes and styrene have been referred to earlier [42]. Odours from combustion processes, such as for example from diesel fuels, are unpleasant. Comparative odour threshold values have been given for benzene of 4.7 ppm, for toluene of 2.1 ppm, for xylene of 0.5 ppm, for chlorobenzene of 0.2 ppm, for styrene of 0.05 ppm, and for nitrobenzene of 0.005 ppm [11]. Slightly lower odour thresholds are reported by other scientists: 0.9–1.6 ppm for benzene, 0.03–0.5 ppm for toluene, and 0.15–0.18 ppm for xylene [12]. In water, quantities higher than 6.8 mg/m^3 naphthalene, 37 mg/m^3 styrene and 140 mg/m^3 ethylbenzene cause unpleasant odours [13]. The following compounds may cause problems: o-dichlorobenzene (max. quantity in surface water $60 \, mg/m^3$, max. quantity in drinking water $1 \, mg/m^3$, odour threshold $2 \, mg/m^3$), p-dichlorobenzene (max. quantity in surface water $200 \, mg/m^3$, max. quantity in drinking water $1 \, mg/ m^3$, odour threshold $30 \, mg/m^3$), trichlorobenzene (max. quantity in surface water $5 \, mg/m^3$, max. quantity in drinking water $0.1 \, mg/m^3$, odour threshold $5 \, mg/m^3$), 1,3,5-trimethylbenzene (max. quantity in surface water $10 \, mg/m^3$, max. quantity in drinking water $1 \, mg/m^3$, odour threshold $20 \, mg/m^3$), naphthalene (max. quantity in surface water $3 \, mg/m^3$, max. quantity in drinking water $0.1 \, mg/m^3$, odour threshold $3 \, mg/m^3$), and 2,6-dimethyl naphthalene (max. quantity in drinking water $1 \, mg/m^3$ and odour threshold $10 \, mg/m^3$) [168]. (For odour data, see also reference [239].)

References

1. Winnacker-Küchler: Chemische Technologie, Vol. 3 (1971)
1a. Collin, G., Zander, M.: Erdöl und Kohle-Erdgas-Petrochem. *33*, 557 (1980)
2. Kirk-Othmer's Encyclopedia of Chemical Technology, Vol. 13 (1967), Interscience Encyclopedia Inc. New York
3. Ullmanns Encyclopädie der Technischen Chemie, 4th Ed. Vol. 8 (1974), Verlag Chemie Weinheim
3a. Stanford Res. Inst. Chem. Econom. Handbooks, 1979
4. Oeldorado 78, Esso, Zürich 1979
5. Korte, F., Boedefeld, E.: Ecotoxicol. Environm. Safety *2*, 55 (1978)
6. Fate and Effects of oil in the sea, Exxon Background Ser. 12 (1978) New York, and Petroleum in the Marine Environment, National Acad. Sci. Washington 1975
7. Esso Magazin 1/78 p. 20, Hamburg
8. Esso-Informations-Programm Nr. 16, Ottokraftstoff, Hamburg
9. Winter, G. (BASF): Swiss Chem. *1*, Nr. 7/8 (1979); Chem. Rundschau *32*, Nr. 39, (26. Sept. 1979)
10. Motorenbenzin und Umwelt, Rep. Swiss Expert Comm. Fed. Departm. Home Affairs, July 1976
11. Moll, W.L.H.: Taschenbuch Umweltschutz, Darmstadt 1973
12. Merian, E.: Chem. Rundschau *27*, Nr. 42, (16. Oct. 1974)
13. Vapor-phase organic Pollutants, Nat. Acad. Sci. Washington 1976
14. Amer. Petroleum Inst., Washington. Personal Communication, Sept. 1979
15. Mineralölwirtschaftsverband, Hamburg, Personal Communication, July 1979
16. Messung u. Ermittlung v. Kohlenwasserstoff-Emissionen, Ber. 16/76, Umweltbundesamt, Berlin 1976
17. Concawe Report 1/73: Benzene in Motor Gasoline, Den Haag 1973
18. Martin, W.: Plan *35*, 14 (1978)
19. Concawe Rep. 3/73: Effect of Gasoline Aromatics Content on Exhaust Emissions, Den Haag 1973
20. Umwelt, Nr. 66 (19. Jan. 1979); U.W.D., 9, Nr. 4 (8. Febr. 1979)
21. Oelert, H.H.: Erdöl u. Kohle, Erdgas Petrochem. *27*, 146 (1974)
22. Oelert, H.H., et al.: Dechema-Kolloq. 1975: Abgasemission individueller Kohlenwasserstoffe aus Ottomotoren (Air Poll. Abstr. 076392)
23. Shinohara et al.: Tokyo Symp. 1975, Fuel Composition and Emission (Air Poll. Abstr. 0766805)
24. Santodonato, J. et al.: Health Effects Associated with Diesel Exhaust, Emissions EPA-600/1-7-063 (1978)
25. Aus der Sprache des Oels, Mineralöl-Wirtschaftsverband e.V. 1975
26. Verkehrstaschenbuch 1976/77, Aral, Bochum, 1976
26a. Huisingh, J. et al.: Application of Bioassay to the Characterization of Diesel Particle Emissions, EPA-600/9-78-027, Sept. 1978
27. Benzene Emission Control Costs in Selected Segments of the Chemical Industry – An Analysis of the F.D. Shell Inc., Florham Park N.J., USA 1978
28. Goethel, G.F.: Achema-Meet., Frankfurt/Main 1979: Sicherer Umgang mit Benzol

29. Shuster, W.W.: Meet. Air Pollution Contr. Assoc. Boston, Mass. 1975: Partial combustion and Pyrolysis of Solid wastes (Air Poll. Abstr. 077704)
30. Bradowsky, T.P., Wilson, N.B., Scott, W.J.: Anal. Chem. *48*, 1812 (1976)
30a. Markt + Betrieb, Chem. Ind. *32*, 81, 106–107, 148–150 (1980)
31. Marsden, C.: Solvent Guide, London 1963
32. Kirk-Othmer's Encyclopedia of Chemical Technology, Vol. 20 (1969), Interscience, New York
33. Kankocho, Kogai et al.: Rep. Comm. Control of Hydrocarbons Sources, Tokyo, Sept. 1976 (Air Poll. Abstr. 100154)
34. Tsuritani et al.: Print J. *57*, 3 (1974): Labor Health and Pollution Control (Air Poll. Abstr. 069282)
35. Basler Zeitg. Nr. 179 (3. Aug. 1979)
36. Winnacker-Küchler: Chemische Technologie, Vol. 4 (1972)
37. Kirk Othmer's Encyclopedia of Chemical Technology, Vol. 22 (1970), Interscience Encyclopedia Inc. New York
38. Ibid. Vol. 19 (1969)
39. Winnacker-Küchler: Chemische Technologie, Vol. 5 (1972)
40. Dörfel, H.: 12th Int. TNO Conf., Rotterdam, 1979, Innovation in the Chemical Industry. Chem. Rundschau *32*, Nr. 13 (28. March 1979)
41. Malotsev, V.V., et al.: Plast. Massy *5*, 35 (1975): Identification of Substances released by Polystyrene Products (Air Poll. Abstr. 079549)
42. Nishikawa, et al.: Imono *46*, 715 (1974): Odors Exhausted in Foundry and its Control (Air Poll. Abstr. 071346)
43. Ullmann's Encyclopädie d. Techn. Chem. 4th Ed., Vol. 9 (1975)
44. Safe, S., Plugge, H.: Chemosphere *6*, 641 (1977)
45. Sumino, K.: Arch. Environ. Contam. Toxicol. *6*, 365 (1977)
46. Roesijadi, G., et al.: Environ. Pollut. *15*, 223 (1978)
47. Malins, D.C., Collier, T.K., Roubal, W.T.: Aquat. Poll. Biol. Eff. *298*, 482 (1977); Arch. Environ. Contam. Toxicol. *5*, 513 (1977); Int. J. Environ. Anal. Chem. *6*, 55 (1979); Chem. Rundschau *32*, Nr. 20 (17. Mai 1978)
48. Neff, J.M., Anderson, J.W.: Bull. Environ. Contam. Toxicol. *14*, 122 (1975); Sources, Effects and Sinks of Hydrocarbons in the Aquatic Environment, Washington, Proc. p. 216 (1976)
49. Varanasi, U., et al.: Arch. Environ. Contam, Toxicol. *8*, 203 (1979); Toxicol. Appl. Pharmacol. *44*, 277 (1978)
50. Schwarz, F.P., et al.: J. Chem. Eng. Data *22*, 270 (1977)
51. Dean-Raymond, D., in: Development in Industrial Microbiology, Vol. 16, Soc. Biol. Sci., Washington D.C. 1975, p. 97
52. Schomburg, G.: Gaschromatographie, Verlag Chemie, Weinheim 1977
53. Chrompack News, no. 29 (1980)
54. Lijinsky, W., Domsky, I.I., Ward, J.: J. Gas Chromatogr. 1965, 152
55. Sauerland, H.-D., Zander, M.: Erdöl u. Kohle-Erdgas-Petrochem. *25*, 526 (1972)
56. Sauerland, H.-D., Zander, M., ibid. *19*, 502 (1966)
57. Grob, K., Grob, G.: J. Chromat. *62*, 1 (1971)
58. D'Orazio, M.: Appl. Spectrosc. *33*, 278 (1979)
59. Hester, N.E., Meyer, R.A.: Environmental Science & Technology *13*, 107 (1979)
60. Swinnerton, J.W., Linnenbom, V.J.: J. Gas Chromatogr. *5*, 570 (1964)
61. Zander, M.: Phosphorimetry – The Application of Phosphorescence to the Analysis of Organic Compounds, Academic Press, New York, London 1968
62. Hornig, A.W.: Internat. Congr. Analyt. Techniques in Environm. Chem., Barcelona, Nov. 27–29, 1978
63. Walther, H.: Informationstagung Spektroskopie, Salzburg, Sept. 17–21, 1979
64. Grob, K.: Neue Zürcher Zeitung, 10. Sept. 1973 (Nr. 419) Aug. 1980
65. Study of Swiss Fed. Inst. for Water Ressources and Water Poll. Control (EAWAG) Dübendorf, 1974
66. Nelson, P.F., et al.: Clean Air J. *11*, 1 (1977)
67. Garrison, A.W. (EPA, Athens, Georgia 30613, USA.) Personal communication
68. Petroff, N. (Inst. du Petrol, F-92502 Rueil Malmaison, France): Tentative Method for Detection of Aromatic Hydrocarbons (C_6 to C_9) in Polluted Water, see Concawe-Rep. Nr. 3/79 (187)

69. Esso-Magazin 3/78, p. 22 (Hamburg 1978), see also Concawe's Assesment of the Environmental Impact of Refinery Effluents, Concawe Report 1/80, Den Haag Febr. 1980 and 5/79, November 1980
70. Yang, W.C., Hsing Wang: Water Res. *11*, 879 (1977)
71. Lu Po-Yung, Metcalf, R.L.: Env. Health Perspect. *10*, 269 (1975)
72. Vogel, H.U.: Chemiker-Kalender, Springer-Verlag 1956
73. CRC-Handbook of Chemistry and Physics, 47th Ed. 1966/1967, CRC-Press, Cleveland, Ohio
74. Timmermans, J., in: Physico-chemical Constants of Pure Organic Compounds, Elsevier, Amsterdam 1950
75. Dictionary of Organic Compounds, London 1965
76. Beilstein, 3. und 4. Ergänzungswerk, Vol. 5, Springer-Verlag, Berlin 1978
77. Mackay, D., Shiu (Dep. Chem. Engin. and Applied Chem., Univ. Toronto, Can.), Personal communications, 1979
78. Gäb, S., et al.: Nature (London) *270*, 331 (1977)
79. Nwanko, J., Amos, T.: Ann. N.Y. Acad. Sci. *237*, 397 (1974)
80. Kasoaka, Sh., et al.: Congr. Proc. Jap. Soc. Chem. Eng. Tokyo *1975*, p. 213
81. Pope, D., Walker, D.S., Moss, R.L.: Atmos. Environ. *10*, 951 (1976)
82. Atkinson, R., Pitts, J.N. Jr.: J. Phys. Chem. *78*, 1780 (1974)
83. Nojima, K., et al.: Ann. Rep. Kanagana Pref. Env. Cent. *6*, 45 (1975)
84. Nojima, K., et al.: Techn. Rept. Air Pollution *17*, 145 (1975)
85. Nojima, K., et al.: Chemosphere *4*, 77 (1975)
86. Nojima, K., et al.: Tech. Rept. Air Pollution *18*, 162 (1976)
87. Nojima, K., Kanno, S.: Chemosphere *6*, 371 (1977)
88. Nojima, K., et al.: Japan Soc. Pharmacy, Tokyo, 1975, p. 109
89. Doyle, G.J., et al.: Environ. Sci. Techn. *9*, 237 (1975)
90. Futsuhara, N., et al.: Int. Clean Air Congr. Proc. *4*, 478 (1977)
91. Akimoto, H., et al.: Conf. Proc. Int. Photochem. Oxidant Poll. Control, Raleigh, II, 737 (1976/1977)
92. Inoue, G., et al.: Conf. Proc. Chem. Society, Japan, Tokyo 34 (1976)
93. Nojima, K., et al.: Techn. Rept. Air Poll. *19*, 128 (1977); Nojima, K., et al.: Chemosphere 5, 25 (1976)
94. Shinohara, H.: Lecture Meet. Japan Soc. Automotive Engineers Science, Tokyo, 1975, p. 101
95. Katou, T.: Bull. Inst. Env. Sci. Technol., Yokohama, *1*, 37 (1974)
96. Shinoyama, E.: J. Japan Soc. Air Poll. *10*, 264 (1975)
97. Chu, R.R.-C.: Univ. Microfilms 74–11, 412 (1974) Ann. Arbor Mich.
98. Sozuki, T. et al.: Lecture Meet. Japan Soc. Automotive Engin. Science, Tokyo 1975, p. 597
99. Lipeles, M., Ratto, J.: Air Poll. Abstr. 104489
100. Kallio, R.E.: Sources, Effects and Sinks of hydrocarbons in the aquatic environment, Washington 1976, p. 218
101. Parker, P.L., et al., in: Sources, Effects and Sinks etc., 1976, p. 257
102. Larson, R.A., et al.: ibid. p. 300
103. Frankenfeld, J.W.: Proc. Joint Conf. Prev. Contr. Oil Spills, 485 (1973)
104. Hoigné, J., Bader, H.: Proc. Intern. Symp. Ozon and Water, GDCh, Berlin 1977, p. 261; Prog. Water Techn. *10*, 657 (1978); Ann. Rep. EAWAG Dübendorf 1977, p. 8
105. Parker, C.A.: AML Rep. B. 198 (M): The Fate of Crude Oil at Sea
106. Petroleum in the Marine Environment, Nat. Acad. Sci. 1975
107. Ivens, G.W.: Ann. Appl. Biol. *39*, 418 (1952)
108. Gibson, D.T. in: Sources, Effects and Sinks etc., 1976, p. 225
109. Gibson, D.T.: Meet. Noordwijkerhout/Amsterdam 1977, Proc. Aquatic Pollutants, 1978, p. 187 (Pergamon; Chem. Rundschau 23, Nov. 1977)
110. Schmidt, H.-L., Schmelz, E.: Chemie in unserer Zeit *14*, 25 (1980)
111. Omai, T., Yamada, K.: Appl. Biol. Chem. *33*, 979 (1969)
112. Davis, R.S., et al.: Canad. J. Microbiology *14*, 1005 (1968)
113. Evans, W.Ch.: Nature (London) *270*, 17 (1977)
114. Dagley, S., Chapman, P.J.: Methods in Microbiol., p. 217, Academic Press 1971
115. Dagley, S.: Essential in Biochemistry *11*, 81 (1975), Academic Press
116. Chapman, P.J., Hopper, D.J.: Biochem. J. *110*, 491 (1968)

117. Hopper, D.J., Chapman, P.J.: ibid. *122*, 19 (1970)
118. Knackmuss, H.-J.: Chemiker-Z. *95*, 213 (1975)
119. McKinney, R.E., et al.: Sewage and Ind. Wastes 28, 547 (1956)
120. Reiner, M., Hegeman, G.D.: Biochemistry *10*, 2530 (1971)
121. Mill, T.: Meeting Noordwijkerhout/Amsterdam 1977, Proc. Aquatic Pollutants, p. 223 (Pergamon 1978); Chem. Rundschau 30, Nr. 47 (23. Nov. 1977)
122. Lilly, M.D.: Achema-Meet. 1979 (Chem. Rundschau 32, Nr. 36, 5. Sept. 1979)
123. Hill, E.C., Cardiff, U.K.: cit. Chem. Rundschau 29, Nr. 42 (13. Oct. 1976): Environmental Research in Great Britain
124. Davies, J.J., Evans, W.C.: Biochem. J. *91*, 251 (1964)
125. Jeffrey, A.M., et al.: Biochemistry *14*, 575 (1975)
126. Cerniglia, C.E., Gibson, D.T.: Appl. Environ. Microbiol. *34*, 363 (1977)
127. Cerniglia, C.E., Hebert, R.L., Szaniszlo, P.J., Gibson, D.T.: Arch. Microbiol. *117*, 135 (1978)
128. Kappeler, Th., Wuhrmann, K., Water Res. *12*, 327, 335 (1978)
129. Schwarzenbach, R.P., et al.: Organic Geochemistry *1*, 93 (1978)
130. Kaspar, H., Leidner, H., Wuhrman, K.: gwf-wasser/abwasser *117*, 400 (1976)
131. Matter-Müller, Chr.A.: Dissertation Nr. 6403, ETH-Zürich (1979)
132. Mara, S.J., Lee, S.S.: Assessment of human exposure to atmospheric Benzene, EPA 450 (1978)
133. Pellizari, E.D., Sawicki, E.: The Measurements of carcinogenic vapors in ambient atmospheres, EPA-600 (1977)
133a. UWD-Report (Düsseldorf) *10*, Nr. 27 (25. Sept. 1980)
134. Mendenhall, D.D., et al.: Organic Characterization of aerosols and vapor phase compounds in Urban Atmospheres, EPA-600/3-78-031 (1978)
134a. Immissionsbelastung durch Kraftfahrzeuge, Umwelt (Bonn) Nr. 77, p. 25 (27. June 1980)
134b. Lahmann, E. et al.: Bundesgesundheitsblatt *21*, 75 (1978)
135. Grob, K.C.: Neue Zürcher Z. 7. Aug. 1972 (Nr. 364)
136. Merian, E.: Chimia *28*, 253 (1974)
137. Air Quality Criteria for Hydrocarbons, 1970, DHEW public.
138. Lonnemann, W.A., et al.: Environ. Sci. Technol. *2*, 1017 (1968)
139. Altshuller, A.P., et al.: Environ. Sci. Technol. *5*, 1009 (1971)
140. Lonnemann, K.A., et al.: ibid. *8*, 229 (1979)
141. Perry, R., et al.: Symp. Environment and Health, Paris 1979
142. Kikuchi, T., Sone, M.: J. Jap. Soc. Air Poll. *9*, 109 (1979)
143. Schewe, G.J., Johnsen, R.J.: EPA internal Rep. 1977
144. Draft Task Rep. Sampling and Analysis, Battelle-Columbus Lab. 1977
145. Hester, N.E., Meyer, R.A.: Environ. Sci. Technol. *13*, 107 (1979)
146. Louw, C.W., et al.: Atmosph. Environm. *11*, 703 (1977)
147. Burghardt, E., Jeltes, R.: Atmosph. Environm. *9*, 935 (1975)
148. Raymond, A., Guiochon, G.: Environ. Sci. Technol. *8*, 143 (1974)
149. Bertsch, W., et al.: J. Chromatogr. Sci. *12*, 175 (1974)
150. Kaci, K.N., et al.: J. Jap. Soc. Air Poll. *9*, 218 (1974)
151. Merian, E.: Chem. Rundschau 29, Nr. 14 and 42, 31. March and 13. Oct. 1976
152. Smith, M.J., et al.: Int. Clean Air Congr. Proc., Australia, 1978, p. 18
153. Altwickler, E.R., et al.: Int. Clean Air Congr. Proc., Tokyo, 1977, p. 520
154. Kopzijnski, S.L., et al.: J. Air Poll Control Ass. *25*, 251 (1975)
155. Blumer. W.: Fortschr. Med. *93*, 1571 (1975)
156. Hirose, K., et al.: J. Jap. Soc. Air Poll. *10*, 586 (1975)
157. Kankocho, K., Senmon, Sh.: Air Poll. Abstr. 100154 (1976)
158. Ciccioli, P., et al.: J. Chromatogr. *126*, 757 (1976)
159. Hasegawa, T., et al.: J. Odor Contr. *6*, 20 (1977) (Air Poll. Abstr. 103855)
160. Braszezijnska, Z.: Prac. Lek. *29*, 182 (1977) (Excerpta Medica)
161. Bobev, G., Cohen, E.: Probl. Hig. Sofia 2, 63, 1976
162. Garrison, A.W.: Consultants Rep. Techn. Paper No. 9 to the WHO Intern. Ref. Centre for Community Water Supply, Dec. 1976
163. Keith, L.H.: Identification and Analysis of Organic Pollutants in Water, Ann. Arbor Sc. Publ. 1976

164. Garrison, A.W.: Meet. Noordwijkerhout/Amsterdam 1977, Proc. Aquatic Pollutants, p. 39 (Pergamon 1978); (Chem. Rundschau 23. Nov. 1977)
165. Kraybill, H.F.: ibid., p. 419; (Chem. Rundschau 23. Nov. 1977)
166. Frentzel-Beyme, R.: Meet. D-6070 Langen, 1979 (Chem. Rundschau 12. Dec. 1979)
167. Malle, K.G.: Chem. in unserer Zeit *12*, 117 (1978)
168. Health Effects relating to direct and indirect Re-use of Waste Water for Human Consumption, WHO-Technical Paper no. 7, Geneva 1975
169. Kuszmaul, H.: Meet. Noordwijkerhout/Amsterdam, 1977, Proc. Aquatic Pollutants, p. 265 (1978)
170. Zoeteman, B.C.J.: ibid., p. 359
171. Giger, W.: ibid., p. 111; Chem. Rundschau 23. Nov. 1977
172. Schwarzenbach, R., et al.: Ann. Report EAWAG, Dübendorf 1978, p. 51
173. Mackay, D.: Meet. Noordwijkerhout/Amsterdam, 1977, Proc. Aquatic Pollutants, p. 175 (1978)
174. Safe, S.: ibid. p. 299 (1978); Chem. Rundschau 30, Nr. 47, 23. Nov. 1977
175. Information System EPA-R2-73-277, Corvallis, USA
176. Information System "Monitoring to Detect Previously Unrecognized Pollutants in Surface Water", EPA 560/6-77-015, Wash. D.C. 1977
177. J. Water Poll. Contr. Fed. (EPA-Information)
178. Environm. Sci. Techn. *9*, 762, 1174 (1975), EPA-Information
179. Information System: Industrial Pollution of the Lower Mississippi River, EPA 3/3/7, Dallas (Texas) 1972
180. Information System: Environmental Appolications of Advanced Instrumental Analyses, EPA FY 73-660/2-74-078
181. Information System: Analytical Report: New Orleans Area Water Supply, EPA 906/9-75-003, Dallas (Texas) 1972
182. J. Chromatography *76*, 45 (1973), Information EPA
183. Filip, Z.: Pres. Meeting, D-6070 Langen, 1979 (Chem. Rundschau 32, Nr. 50, 12. Dec. 1979)
184. Matthess, G.: ibid.
185. Sontheimer, H.: ibid.
186. Wolters, N., von Kunowski, J.: ibid.
187. Concawe Rep. 3/79: Protection of Groundwater from Oil Pollution, Den Haag, Netherlands, 1979
188. Perry, R.; cit. Chem. Rundschau 29, Nr. 42, 13. Oct. 1976: Environm. Res. in Great Britain
189. Knausenberger, H.: Pres. Meeting, D-6070 Langen, 1979 (Chem. Rundschau 32, Nr. 50, 12. Dec. 1979)
190. Boyd, B.D., et al.: Congr. Proc. Sources, Effects and Sinks of Hydrocarbon in the Aquatic Environment, Washington, 1976, p. 38
191. Brown, R.A., Searl, T.D.: ibid., p. 240
192. The Separation of Oil from Water for North-Sea Oil Operations, Poll. Papers Nr. 6, Centr. Unit. Envir. Poll. Dep. of the Environment, London (1976)
193. Zsolnay, A., et al.: Environ. Conserv. *5*, 295 (1978)
194. Mazliak, P.: Ann. Nutr. Aliment. *28*, 277 (1974)
195. Bauer, D.: Meet. Achema, 1979 (Chem. Rundschau 32, Nr. 41, 10. Oct. 1979)
196. Sterner, J.H., et al.: Assessment of Health Effects of Benzene Germane to low-level Exposure, EPA-600/1, 1978
197. Fodor, G.G.: Schädliche Dämpfe, VDI-Verlag, Düsseldorf 1972
198. Benzol am Arbeitsplatz, DFG-Arbeitsgruppe (G. Büttner and D. Henschler), H. Boldt-Verlag, Boppard/Verlag Chemie, Weinheim 1974
199. Human and Animal Toxicology of Benzene, Rep. of CISHEC/IP Joint Working Party (H.G.S. van Raalte et al.), Shell, Den Haag 1974
200. Nomiyama, K., and H.: Int. Arch. Arbeitsmed. *32*, 85 (1974)
201. Nomiyama, K., and H.: ibid. *32*, 75 (1974)
202. Gut, I.: Europ. Soc, Tox. Dresden, cit. Chem. Rundschau 32, Nr. 35 (1979)
203. Förster (Frankfurt) and Linder (Hamburg) Chem. Rundschau 29, Nr. 14 (31. March 1976): Environmental Research in Germany
204. Lehmann, H. (Shell): Erdöl und Kohle-Erdgas-Petrochemie *32*, 331 (1979)
205. Berlin, M., et al.: Meet. Europ. Soc. Toxicol. Berlin 1978 (Chem. Rundschau 31, Nr. 34, 23. Aug. 1978)
206. Sato, A., et al.: Brit. J. Ind. Med. *32*, 321 (1975) (Medline)

207. Radojiac, B.: Arch. Hig. Rada Toksikol. *26*, 209 (1975)
208. Gig. Truda: Zabol Moskwa 1979, Nr. 2; cit. Medizin in Osteuropa, Freie Universität Berlin, 10. Aug. 1979; U.W.D.-Umweltschutzdienst, Düsseldorf, 9, Nr. 24, 29. Aug. 1979
209. Tokunaga, R., et al.: Int. Arch. Arbeitsmed. *33*, 257 (1974)
210. Atkinson, L.P., et al.: Water Air Soil Poll. *8*, 235 (1977)
211. Truhout, R.: Meet. Secotox, Vienna, 1978 (Chem. Rundschau 31, Nr. 47, 22. Nov. 1978)
212. Walsh, G.E., Ainsworth, K.A., Faas, L.: Bull. Environ. Contam. Toxicol. *18*, 297 (1977)
213. Melancon, M.J. Jr., Lech, J.J.: Arch. Environ. Contam. Toxicol. *7*, 207 (1978)
214. Bakke, T., Skjoldal, H.R.: Mar. Poll. Bull. *10*, 111 (1979)
215. Oesch, F., Biozentrum CH-4056 Basel, Personal Communication, 1974
216. Oesch, F.: Meet. Europ. Soc. Toxicol., Berlin, 1978, cit. Chem. Rundschau 31, Nr. 34, 23. Aug. 1978
217. Jollow, D.J., et al.: Workshop: Quantitative Aspects of Risk Assessment in Chemical Carcinogenesis, Rome, Italy, 1979 (Chem. Rundschau 32. Nr. 19, 9. May 1979)
218. Harper, C., et al.: Drug Metab. Disposition *3*, 381 (1975)
219. Snijder, R., Gonasn, L., et al.: Toxicol. Appl. Pharmacol. *11*, 346 (1967); *26*, 398 (1973)
219a.Drew, R.T., Fouts, J.R.: ibid. *27*, 183 (1974)
220. Kocsis, J.J.: Lecture Guildford, Surrey, U.K., Sept. 1979; Chem. Rundschau 33, Nr. 4, 23. Jan. 1980
221. Nakamura, S., Nakajima, T.: Ind. Health *5*, 238 (1967)
222. Gerarde, H.W., Ahlstrom, D.A.: Toxicol. Appl. Pharmacol. *9*, 185 (1966)
223. Jerina, D.M., et al.: Biochemistry *9*, 147 (1970)
224. Conning, D., Gordon, M.: Symp. Toxicol., Surrey Univ., Guildford, U.K., 1979 (Chem. Rundschau 33, Nr. 4, 23. Jan. 1980)
225. Luchi, L.D., Bath, L.J.: Toxicol. Environ. Health (Suppl.) *2*, 107 (1977)
226. Goldstein, B.D.: ibid. *2*, 1 (1977)
227. Haley, F.J.: Clin. Toxicol. *11*, 531 (1977)
228. Albert, R.E., et al.: Carcinogen Assessment Group's Final Rep. on Population Risk to ambient Benzene Exposure, EPA 1978
229. Tough, J.M., Smith, P.G.: Europ. J. Cancer *6*, 49 (1970)
229a.Kala, A. et al.: Bull. Env. Contam. Toxic. *1978*, 287
230. Kilian, D.J., et al.: A Cytogenic study of workers exposed to Benzene in the Texas Division of Dow Chemical, USA, 1978
231. Picciano, D.: Communication submitted to EPA, 1978
232. Saito, K., Takakuwa, E.: Jap. J. Ind. Health *16*, 3 (1974) (Air. Poll. Abstr. 066891)
233. Borneff, J., et al.: Umweltforschung in d. Bundesrepublik (Chem. Rundschau 29, Nr. 14, 31. March 1976)
234. Stöfen, D.: Luftreinhaltung, Umweltmagaz. 34 (1978)
235. Blumer, W.: Forschr. Med. *93*, 1571 (1975)
236. Overman, O.: Meet. Achema, Frankfurt/M. 1979 (Chem. Rundschau 32, Nr. 41, 10. Oct. 1979)
236a.Merian, E.: Chemosphere *9*, Nr. 5/6 (June 1980)
237. Schunk, W. et al.: Z. Ärztl. Fortbild. *71*, 932 (1977)
238. Goodman, S., Gilman, A.: The Pharmacological Basis of Therapeutics, 1970, p. 930
239. Verschueren, R.: Handbook of Environmental Data on Organic Chemicals, Van Nostrand, New York 1977
240. Snijder, R., Kocsis, J.J.: Curr. Concepts of Chronic Benzene Toxicology, CRC crit. Rev. Toxicol. *3*, 265 (1975)
241. Truhaut, R., et al.: Int. Workshop on Toxicology of Benzene, Paris, 1976; Int. Arch. Occ. Environ. Health *41*, 65 (1978)
242. Osawa, S.: Air Poll. Abstr. 84697
243. Wildman, J.M. et al.: Comm. Chem. Pathol. Pharm. *13*, 473 (1976)
244. Sato, A., et al.: Brit. J. Ind. Med. (London) *32*, 210 (1975)
245. Uyeki, E.M., et al.: Toxicol. Appl. Pharmacol. *40*, 49 (1977)
246. Khan, H., Khan, M.H.: Arch. Toxicol. *31*, 39 (1973)
247. Speck, B.: Haematol. Bluttransfus. *16*, 235 (1975)
248. Greenblatt, D.R. et al.: Environ. Res. *13*, 425 (1977)
249. Cassan, G., Baron, J.: Arch. Malad. Prof. *17*, 602 (1956)

250. Gallinelli, R.: Med. Lavoro *57*, 257 (1966) (Zbl. Arbeitsmed. *18*, 280 (1968))
251. Sram, R.J.: Meet. Europ. Soc. Tox., Dresden, 1979 (Chem. Rundschau 32, Nr. 35, 29. Aug. 1979)
252. Reske: UFOKAT PD-047 (Chem. Rundschau 29, Nr. 14, 31. March 1976)
253. Stieglitz, R., et al.: Arch. Geschwulstforsch. *44*, 145 (1974)
254. Brandt, L., et al.: Lancet *2*, 1074 (1977)
255. Tabershaw, I.R., Lamm, S.H.: Lancet *2*, 867 (1977)
256. Snijder, R., et al.: Life Sci. *21*, 1709 (1977)
257. Infante, R.F., et al.: Lancet *2*, 76 (1977); *2*, 867 (1977)
258. Askoy, M., et al.: Am. J. Med. *52*, 160 (1972); Blut *28*, 293 (1974); Blood *44*, 837 (1974); Acta Haemat. *55*, 65 (1976); Testimony to Occ. Saf. Health Adm., US Dep. Labor, 1977; Blood *52*, 285 (1978)
259. Ott, M.G., et al.: OSHA Benzene Hearings, July–Aug. 1977; Arch. Environ. Health *33*, 3 (1978)
260. Dean, B.J.: Mutat. Res. *47*, 75 (1978)
261. Schlipköter, H.W.: Private Communication; E.C. Commission for Noise Abatement and for Purification of the Atmosphere, 1974, p. 203
262. Cohen, H.S., et al.: Am. J. Med. Sci. *275*, 124 (1978)
263. Hara, E.I.: Jap. J. Ind. Health *3*, 231 (1961)
264. Novikov, Y.U.: Gig. I. Sanit. *21*, 20 (1956); Volkora, A.P.: ibid. *24*, 80 (1959); Gusey, J.S.: ibid. *30*, 331 (1965); *32*, 159 (1967); Gotmelker, V.H.: ibid. *33*, 327 (1968); *33*, 122 (1968); Karkov, A.P.: ibid. *37*, 2 (1972)
265. Bokina, A.I., et al.: Environ. Health Perspect *13*, 37 (1976)
266. Med. Biol. Lab., TNO Environ, Protection Dep. Rijswijk, The Netherlands (Chem. Rundschau 29, Nr. 14, 31. March 1976: Research in the Netherlands)
267. Rothe, A.: Z. Ärztl. Fortb., Jena, *66*, 758 (1972)
268. Bell, A.: Med. J. Australia, *1*, 817 (1957); ref. 12
269. Selikoff, I.J.: IARC Scient. Publ. Nr. 16, 247 (1977)
270. Withey, R.J., Hall, J.W.: Toxicology *4*, 5 (1975)
271. Aleksandrowicz, J. et al.: Med. Progr. *28*, 453 (1977)
272. Verzilora, O.V., et al.: Gig. I. Sanit. *8*, 20 (1978)
273. Gardner, D.E., et al.: J. Toxicol. Environ. Health *3*, 811 (1977)
274. Inoue, K.: J. Osaka City Med. Center *24*, 783 (1975) (Excerpta Medica)
275. Bobey, G., Cohen, E.: Probl. Hig. (Sofia) *2*, 63 (1976) (Excerpta Medica)
276. Matsushita, T., et al.: Ind. Health Japan *13*, 115 (1975) (Air Pollution Abstr. 083489)
277. Smolik, R., et al.: Ind. Arch. Arbeitsmed. *31*, 243 (1973)
278. Ungvary, G., Ankara, H., et al.: Morphol. Igazsagugyi Orv. Sz. *15*, 209 (1975) (Medline)
279. De Rosa, E., et al.: Lav. Um. (Italia) *26*, 144 (1974)
280. Takeuchi, Y., Hisanaga, N.: Brit. J. Ind. Med. *34*, 314 (1977)
280a.Andersson, K. et al.: Toxicology Letters *5*, 393 (1980)
281. Schädigung durch Xylol, Münch. Med. Wochenschr. *114*, 1302 (1972)
282. Clark, W.E.: Am. J. Clin. Pathol. *68*, 425 (1977)
283. Mlynarczijk, W., et al.: Med. Pract. *28*, 243 (1977) (Medline)
284. Liublina, E.J., et al.: Gig. Tr. Prof. Zabol. *5*, 27 (1977) (Medline)
285. Carpenter, C.P., et al.: Toxicol. Appl. Pharmacol. *33*, 543 (1975)
286. Ashan, G., et al.: Acta Otolaryngol (Stockh.) *84*, 370 (1977)
287. Altman, A.T.: Arch. Dermatol. *113*, 1460 (1977)
288. Hernberg, Sv. (Helsinki): 2nd Intern. Congr. Toxicol. 1980, Brussels
289. Seppaleinen, A.M. (Helsinki): 2nd Intern. Congr. Toxicol. 1980, Brussels, cit. for instance by Merian, E.: Chemische Rundschau *33*, Nr. 34, p. 5 (20th August 1980)
290. Garattini, E., et al.: Experentia *37*, 230 (March 1981)

Annex

When this chapter was just typeset, two important papers on environmental fate were published:

291. Kluge, A., Korte, F. and Klein, W., "Benzene in the Environment", Workshop 1980 in Vienna, Austria, Interpretation Data and Evaluation of Current Knowledge, Minutes of the Forum für Wissenschaft, Wirtschaft und Politik e. V., Bonn Reports, Volume 9, March 1981: Benzene evaporates very easily from water, and its half life in water is only about 37 min at room temperature. The time of 50% mineralization in the atmosphere is likely to be about two days. Profile analytical tests show that benzene is not accumulated to a critical extent in algae and in fish. Benzene is also a natural constituent of foodstuffs of plant and animal origin.

292. Calamari, D.A., Presentation "Ecotoxicological Profile of p-Dichlorobenzene" and "Evaluating the Hazard of Organic Substances on Aquatic Life: The para-Dichlorobenzene Example", Symposium "The Scientific Basis for the Assessment of Hazards from Chemicals Associated with the Environment", Imperial College, London September 1981, Ecotoxicology and Environmental Safety (Academic Press) 6 (1982): Calamari, D.A. studied as a model the environmental fate of p-dichlorobenzene, which is produced globally in amounts not less than 80 000 t annually. The half-life in a turbulant river of 1 m depth is about one day, and the distribution between water, soil and air adjusts to about 1.3%, resp. 1.2%, resp. 97.5%, especially since p-dichlorobenzene is relatively volatile and only little water soluble. Rhine water contains about 0.004 mg/l, and a water quality standard of 0.05 mg/l is suggested, mainly because the compound accumulates more in fish hatching than in adult organisms. Acute toxicity levels for algae, daphnia and fish are in the order of 0.8–4.2 mg/l. Fertility of daphnia may be reduced at 0.4 mg/l, whereas no chronical toxicity for fish was discovered at less than 0.1 mg/l. The accumulation in the air could be more critical than the concentrations found in water.

Surfactants

K. J. Bock, H. Stache

Chemische Werke Hüls AG, P. O. Box 1320
D-4370 Marl, Federal Republic of Germany

Chemistry

Introduction and Historical Review

From ancient times, highly developed and civilised people, for example the Romans, have used cleaning agents of animal origin in the form of putrefied urine. Natural cleaning agents, such as soapwort, horse chestnut, wood ash and the like, were used until the Middle Ages. In parallel, soap developed as a very ancient chemical product, from origins which go back to the Sumerians and to Egypt. In the Euphrates and Tigris area, clay tablets have been found, and also ancient papyri, which describe recipes for the preparation of soap-like cleaning agents from animal and vegetable fats and oils, by means of natural soda and potash [1]. At that time, there were certainly no problems regarding toxicology, fish toxicity or biological degradation.

In the course of time, the preparation of soap was perfected, and developed into an art, by the soapboilers' guilds [2]. For many centuries, soap as a cleaning agent for the body, for clothes and for utensils has met the demands for hygiene and cleanliness. Environmental problems, from today's point of view, were ignored at that time. The dilution factor in the effluent was much higher than today and thus the concentration of pollutants in rivers and lakes was substantially lower. The residual soaps in the effluent were either precipitated by the calcium and magnesium salts in the water, or the soap was directly biodegraded by microorganisms.

With the onset of the industrial era, the population density and hence the pollution of surface waters increased. This development in the densely populated industrial areas entailed problems in water management and hygiene. In the overcrowded centres, it became necessary to supply all these people with satisfactory drinking water and service water, and systematically to dispose of the effluents, and to purify them, by removing the diverse waste materials. Thanks to this fact was it possible to banish the spectre of the plague which had decimated whole towns in the Middle Ages. At the same time, new washing methods and explicit bodily hygiene were developed, new fibres and fabrics came into use and new raw materials were discovered.

This development towards new surface-active substances started approximately in the first half of the last century, when vegetable oils were rendered water-soluble by treatment with sulfuric acid. The sulfate formed by the reaction with ricinoleic acid, as the so-called Turkey red oil, replaced the oily mordant, used until then, obtained from olive oil, oxblood and sheep's dung. Even today, Turkey red oils are used as dyeing oils and textile oils [1].

A further milestone was the development of the fatty alcohol sulfates (the sulfuric acid half-esters of stearyl alcohol and oleyl alcohol), which were readily obtainable from sperm oils. A significant advance in the use of the fatty alcohol sulfates, however, occurred only after it had become possible to obtain fatty alcohols of any desired C number from fats or fatty acids by catalytic high-pressure hydrogenation [1]. The first household detergent based on a native raw material converted in this way, in the Federal Republic of Germany, was the FEWA product in 1932 [3]. The direct sulfonation of hydrocarbons to give secondary alkanesulfonates, the incorporation of aromatics into the molecule to give alkylnaphthalenesulfonates or alkylbenzenesulfonates, demonstrated completely new routes. This meant turning away from the surfactant chemistry of native raw materials in favour of synthetic chemistry based on coal and crude oil as the raw materials. In this phase of development, the term syndets (= synthetic detergents) began to be used.

Nature of Surfactants

The term surfactants covers water-soluble, surface-active compounds which are used for wetting, washing, emulsifying and dispersing. Surfactants are characterised in that they concentrate at surfaces and reduce the surface tension. The term tensides, the German word for surfactants, is derived from the Latin: It is based on the root "tensio". A prerequisite for this surface activity is an asymmetric structure of the surfactant molecule which consists of a water-repellent = hydrophobic and a water-attracting = hydrophilic part [4].

In soap, the hydrophobic group is a relatively long aliphatic hydrocarbon radical; the carboxyl group, characteristic of acids and neutralised with an alkali, acts as the hydrophilic group. In order to obtain new surfactants with improved properties, these two groups are varied, at first empirically and then systematically. The hydrophilic and hydrophobic parts of the surfactant molecule are in a balanced mutual relationship. Depending on the molecular structure, the character of the

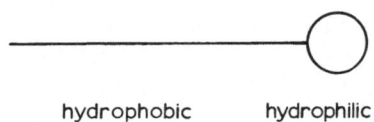

hydrophobic   hydrophilic

Fig. 1. Pattern of a surfactant molecule

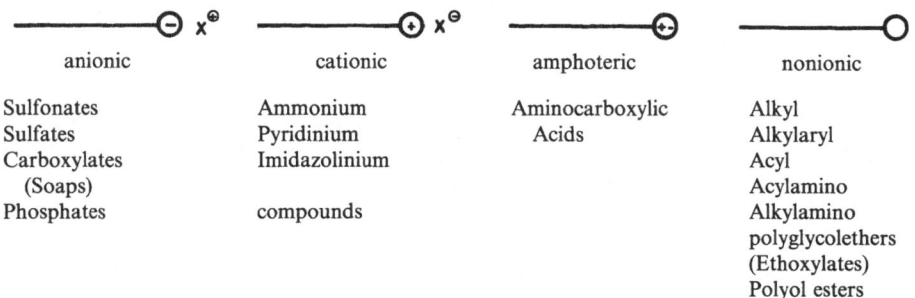

anionic	cationic	amphoteric	nonionic
Sulfonates	Ammonium	Aminocarboxylic	Alkyl
Sulfates	Pyridinium	Acids	Alkylaryl
Carboxylates	Imidazolinium		Acyl
(Soaps)			Acylamino
Phosphates	compounds		Alkylamino
			polyglycolethers
			(Ethoxylates)
			Polyol esters

Fig. 2. Types of surfactants

surfactants is salt-like or not salt-like, and they can be sub-divided into ionic, nonionic, cationic and amphoteric products [5]. As shown in Fig. 2, the hydrophilic or polar group of the anionic surfactants consists in most cases of a sulfonate, sulfate or carboxyl group. In nonionic surfactants, the hydrophilic character is provided in most cases by a polyglycol ether grouping. Cationic surfactants contain, almost without exception, a quaternary ammonium ion as the hydrophilic group. Finally, amphoteric surfactants are compounds having electropositive and electronegative parts in the molecule and, depending on the pH range, they can act as anionic or cationic surfactants. The hydrophobic component, on which all the surfactant types are based, is in most cases a relatively long alkyl chain or alkylaryl chain.

Amongst the products listed, anionic surfactants still represent the quantitatively most important group today (about 60%–70%) in the consumption in the Federal Republic of Germany, whilst the nonionic compounds have a share of about 30%. The latter, however, are assumed to have a higher growth rate, since these compounds display technological properties which are superior in the current modern synthetics. At about 10%, the cationic surfactants and amphoteric products have only a relatively minor, though interesting, position.

Surfactant Classes; Syntheses

Anionics

Alkylbenzenesulfonate (LAS)

Amongst the anionic surfactants, alkylbenzenesulfonate, the industrial manufacture of which was developed in Germany and the USA after 1930, is produced in the largest quantities [6]. Its manufacture is an example of the influence which ecological and legislative aspects have exerted on the development. Initially, the alkylbenzene required for the manufacturing process was obtained by monochlorination of kogasin [hydrocarbons from the Fischer-Tropsch synthesis] fractions or petroleum fractions and a subsequent Friedel-Crafts reaction with benzene [7]. When kogasin fractions were no longer available after the Second World War, tetrapropylenebenzenesulfonate was produced from tetramerised propylene.

With increasing consumption of the alkylbenzene, the branched alkyl chain proved to be troublesome, since the surfactants obtained from this alkylbenzene were relatively resistant biologically. On the other hand, alkylbenzenesulfonates having unbranched alkyl chains were degraded to an extent of 95% or more. Attempts were made to discover, and to manufacture, a suitable straight-chain raw material for the manufacture of biodegradable alkylbenzenesulfonates.

The first possible solution was to subject straight-chain waxy paraffins to thermal cracking and to use the α-olefines thus formed as the alkylation component.

This process can be operated economically only if sufficient crude oil containing this waxy paraffin is available in a refinery, in the course of petroleum refining.

A further process which is of interest and which is the main process operated in practice today, is the synthesis of olefines via straight-chain paraffins. The so-called molecular sieve processes, which became known after 1960, are used for this synthesis [8]. The required n-paraffins of the desired chain length occur in the kerosene fractions of petroleum in a quantity of 20%–30%, on the average. They are, however, mixed with highly branched and cyclic paraffins and aromatics, which must be removed.

There are several processes for producing the olefines or chloroparaffins required for the further reaction with benzene from these n-paraffins. The classical process is the clorination of paraffins to give chloroparaffins and the further reaction with AlCl$_3$ according to Friedel-Crafts [9].

More recent processes use olefine mixtures which are obtained either by thermal dehydrogenation or by chlorination and dehydrochlorination of paraffins [10]. The subsequent alkylation can be carried out not only with AlCl$_3$, but also with HF [11, 12].

Even though a number of relationships between the molecular structure and the technological properties were known from the investigations carried out over many years with tetrapropylenebenzenesulfonate, it was necessary to carry out a new optimisation in view of the new ecological conditions.

The main subject of investigation was the influence of the unbranched alkyl radicals on various parameters. It will be seen from Fig. 4 that the optimum of the detergency, that is to say the cleaning action on textiles, of the biodegradability and of the effect on fish lies in the C_{10}–C_{13} range [13].

With increasing chain length, the products become more harmful to fish. The sensitivity to hardness in the water also rises with increasing chain length.

This development was prescribed by the Detergent Law which came into force in the Federal Republic of Germany in 1961 and which demanded a certain biodegradation rate of the anionic surfactants marketed [14]. This was probably the first time that a development was not initiated by economic or technical problems; there were also no additional hygienic or technological demands to be met. This large-scale industrial development was governed by the ecological aspects.

In the same way as the development of the synthetic surfactants was determined by requirements relating to the environment, as shown by the example of alkylbenzenesulfonate, other examples in the production of surfactants also show how the industry has adjusted to problems of environmental pollution.

The best conditions for minimum pollution of the environment are provided by a combination of effluent purification in the production plant with final puri-

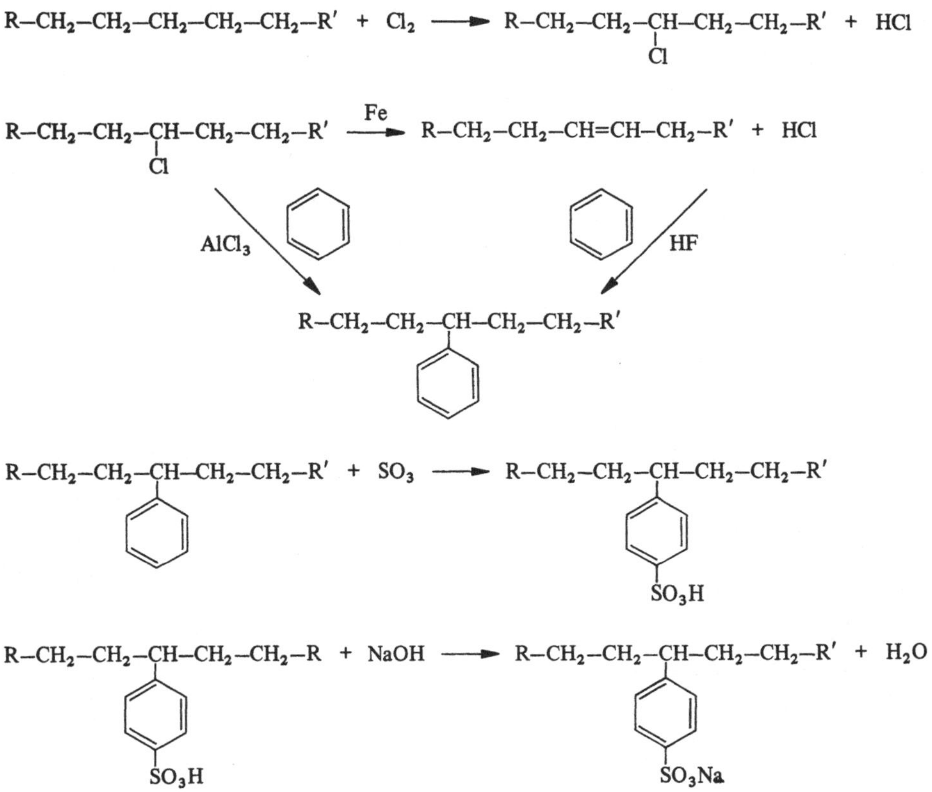

R–CH$_2$–CH$_2$–CH$_2$–CH$_2$–CH$_2$–R′ + Cl$_2$ $\longrightarrow$ R–CH$_2$–CH$_2$–CH–CH$_2$–CH$_2$–R′ + HCl
$\quad\quad\quad\quad\quad\quad\quad\quad\quad\quad\quad\quad\quad\quad\quad\quad\quad\quad\quad\quad\quad$ |
$\quad\quad\quad\quad\quad\quad\quad\quad\quad\quad\quad\quad\quad\quad\quad\quad\quad\quad\quad\quad\quad$ Cl

R–CH$_2$–CH$_2$–CH–CH$_2$–CH$_2$–R′ $\xrightarrow{\text{Fe}}$ R–CH$_2$–CH$_2$–CH=CH–CH$_2$–R′ + HCl
$\quad\quad\quad\quad$ |
$\quad\quad\quad\quad$ Cl

AlCl$_3$ (benzene) $\quad$ (benzene) HF

R–CH$_2$–CH$_2$–CH–CH$_2$–CH$_2$–R′
(phenyl)

R–CH$_2$–CH$_2$–CH–CH$_2$–CH$_2$–R′ + SO$_3$ $\longrightarrow$ R–CH$_2$–CH$_2$–CH–CH$_2$–CH$_2$–R′
(phenyl) $\quad\quad\quad\quad\quad\quad\quad\quad\quad\quad\quad\quad\quad\quad\quad\quad\quad\quad$ (phenyl–SO$_3$H)

R–CH$_2$–CH$_2$–CH–CH$_2$–CH$_2$–R + NaOH $\longrightarrow$ R–CH$_2$–CH$_2$–CH–CH$_2$–CH$_2$–R′ + H$_2$O
(phenyl–SO$_3$H) $\quad\quad\quad\quad\quad\quad\quad\quad\quad\quad\quad\quad\quad\quad\quad\quad\quad$ (phenyl–SO$_3$Na)

Fig. 3. Routes to linear alkylbenzene: chlorination, dehydrochlorination and alkylation with Friedel-Crafts catalysts and subsequent sulfonation to alkylbenzenesulfonate

fication in central primary and biological treatment plants, before discharge to the receiving waters. To meet these conditions, however, it is essential not to allow the residual products obtained during production to become "wastes", but to utilise them as quantitatively as possible in the production cycle. An example of such a combined system within the factory (recycling) is the formation of hydrogen chloride in the production of LAS. The chlorinated hydrocarbons and olefines required for the abovementioned Friedel-Crafts reaction are produced by chlorination or chlorination/dehydrochlorination.

In both cases, gaseous dry hydrogen chloride is formed which can be utilised directly, without expensive purification, for the production of vinyl chloride or polyvinyl chloride. The production of surfactants has here been logically combined with the production of plastics [18] (Fig. 5).

The alkylation can also be carried out by processes which do not pollute the environment. As already stated, the conventional process is operated with AlCl$_3$ catalysis, and the NaCl formed in the course of production, the aluminium hydroxide sludge and residual benzene must be removed before the effluent is discharged. In the alkylation with HF, virtually no waste appears in the effluent, since the HF

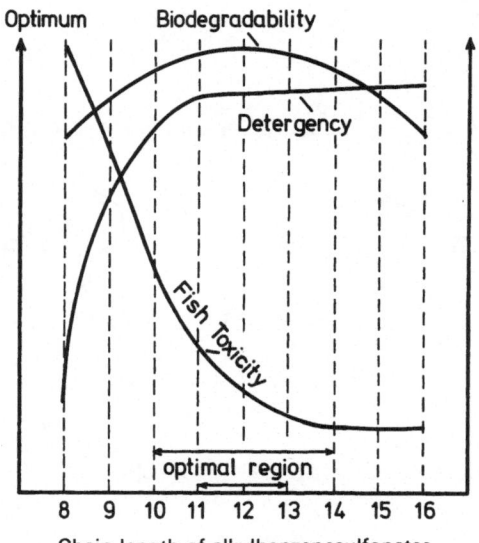

Chain length of alkylbenzenesulfonates

Fig. 4. Detergency, biodegradability and fish toxicity of linear alkylbenzenesulfonates (LAS)

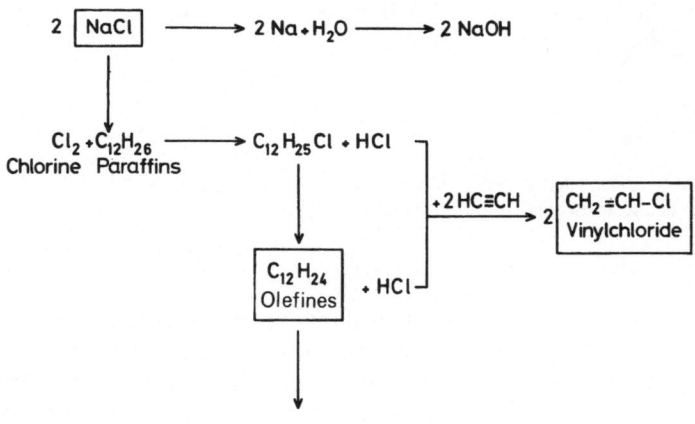

to Alkylation of Benzene

Fig. 5. Olefine synthesis combined with the production of vinyl chloride

is returned into the reaction circulation; the amount consumed, about 0.2 kg per 100 kg of alkylbenzene, passes with the off-gas through an alkali wash, is precipitated as CaF_2 and is deposited as a solid. More recent investigations make it possible to recycle this CaF_2 into the production of HF.

The sulfonation of the alkylbenzene with SO_3 to give the sulfonic acid is another example of an industrial process which does not pollute the environment. When oleum or sulfuric acid is used in the process, a waste sulfuric acid is obtained, which raises disposal problems. This waste acid can be reconverted to concentrated sulfuric acid or SO_3 by a thermal cracking process [15].

Various Anionics

Whilst the alkylbenzenesulfonates are equally used for powder and liquid detergents and cleaners, the field of use of the secondary alkanesulfonates (also called paraffin sulfonates) is rather in the liquid cleaners sector, because of their high solubility. This type of surfactant, produced by the conventional sulfochlorination process, became of some importance as early as the Second World War.

Today, the sulfoxidation process is preferred, because on the one hand there is no use of chlorine, and on the other hand the formation of NaCl is avoided. Like the sulfochlorination process, the sulfoxidation process is carried out with SO_2, under the action of light.

At approximately the same time, a further process for the manufacture of anionic surfactants of the sulfonate type, namely the olefinesulfonates, became of interest. This process starts from α-olefines obtained by cracking of higher hydrocarbons. In this process, the olefine is reacted with SO_3 for a short time in a thin-layer reactor. Alkenesulfonates and hydroxyalkanesulfonates are formed [18].

Shell has developed a new process which leads from ethylene to higher α-olefines (SHOP process, which means Shell-higher-olefines-process) [19].

In addition to these sulfonates which represent the main proportion of the total production of anionic surfactants, other special products are very important as raw materials for cosmetics, textile aids, light-duty detergents and the like.

Table 1 shows the most important anionic surfactants.

Table 1. Classification of anionic surfactants

Type	Example	Formula
Soap	Sodium stearate	$CH_3(CH_2)_{16}$—$COO^{\ominus}Na^{\oplus}$
Sulfonated aromatic hydrocarbons	Linear alkylbenzenesulfonate	$C_{12}H_{25}$—$\langle\!\!\bigcirc\!\!\rangle$ $SO_3^{\ominus}Na^{\oplus}$
Sulfonated aliphatic hydrocarbons	Sec. alkanesulfonate, paraffinsulfonate	$C_{15}H_{31}SO_3^{\ominus}Na^{\oplus}$
Sulfonated α-olefine	α-Olefinesulfonate	$C_{16}H_{31}SO_3^{\ominus}Na^{\oplus}$
Sulfated fatty alcohol	Sodium laurylsulfate	$C_{12}H_{25}OSO_3^{\ominus}Na^{\oplus}$
Sulfated fatty alcohol Ethoxylate	Sodium lauryl polyglycol Ether-sulfate	$C_{12}H_{25}(OCH_2CH_2)_3OSO_3^{\ominus}Na^{\oplus}$
Sulfonated fatty acid Methyl ester	Palm-kernel oil sulfo-fatty Acid methyl ester	$C_{16}H_{33}CH$—$COOCH_3$ $\underset{\displaystyle SO_3^{\ominus}Na^{\oplus}}{\mid}$
Sulfonated maleic ester	Laurylsulfosuccinate	ROOC—CH—CH_2—COOR $\underset{\displaystyle SO_3^{\ominus}Na^{\oplus}}{\mid}$
Carboxymethylated fatty Alcohol ethoxylates	Lauryl polyglycol ether-acetate	$C_{12}H_{25}(OCH_2CH_2)_nOCH_2COO^{\ominus}Na^{\oplus}$
Phosphated alcohol	Mixture of alkyl phosphate and dialkyl phosphate	$ROPO_3^{\ominus}Na^{\oplus}/(RO)_2PO_2^{\ominus}Na^{\oplus}$

Cationics

Even though cationic surfactants play only a subordinate role in the pollution of the environment, they are discussed here for the sake of completeness. Cationic surfactants differ from the anionic surfactants in that the ion carrying the surfactant character is positively charged.

$$\left[\begin{array}{c} R^3 \\ | \\ R^1 - {}^\oplus N - R^4 \\ | \\ R^2 \end{array} \right] X^\ominus \qquad \left[\begin{array}{c} \\ \end{array} \right] X^\ominus \qquad \left[\begin{array}{c} \\ \end{array} \right] X^\ominus$$

Cationic surfactants gained importance, when their bacteriostatic properties were discovered [20]. They are now used as disinfectant and antiseptic components in cosmetic formulations and in medicine. Further uses were discovered after 1950, for example as antistatic agents, textile softeners, corrosion inhibitors, anti-foams, flotation agents and additives for asphalt and petroleum.

In spite of their wide range of uses, their consumption is only 5% of the total surfactant use.

A property common to the cationic surfactants is that their positively charged surface-active part is readily absorbed by the substrate which, in most cases, is negatively charged in water. This is the reason for their bactericidal action in disinfection and their conditioning effect on textile fibres.

Nonionics

Alkyl and Alkylaryl Polyglycol Ethers

The nonionic surfactants, also called nonionics, are almost exclusively addition compounds of ethylene oxide or propylene oxide, that is to say they are substituted polyglycol ethers.

Ethylene oxide is produced by direct oxidation of ethylene with atmospheric oxygen over silver oxide catalysts.

In the conventional chlorohydrin process, which is little used now, relatively large quantities of reaction water polluted with organic substances are obtained. Moreover, the disposal of the alkali metal chloride formed raises problems. Undesirable by-products are almost completely eliminated in direct oxidation. Glycol-containing effluents are recycled into the production sequence.

Ethylene oxide reacts readily with all H-acid compounds, for example an alkylphenol, fatty alcohol, fatty acid, fatty acid amide, fatty amine, mercaptan, glycol and polyglycol (Table 2).

Propylene oxide undergoes the same reactions as ethylene oxide.

In terms of quantities, the alkyl polyglycol ethers or fatty alcohol polyglycol ethers, which are readily obtainable by reaction of ethylene oxide with natural or synthetic alcohols, are of particular interest. For a long time, natural fatty alcohols, which can easily be prepared by hydrogenation of the corresponding fatty acids or methyl esters, were the main raw material base [21]. Neither the manufacture nor

Table 2. Important nonionic surfactants

General formula: $RXH + nCH_2\!\!-\!\!CH_2 \rightarrow RX(CH_2CH_2O)_nH$ $n = 1-50$

(with epoxide O bridging the CH_2—CH_2)

Type	Formula
Fatty alcohol ethoxylate	$CH_3(CH_2)_mO(CH_2CH_2O)_nH$
Alkylphenol ethoxylate	$R\!-\!\langle\bigcirc\rangle\!-\!O(CH_2CH_2O)_nH$
Fatty acid ethoxylate	$R\!-\!COO(CH_2CH_2O)_nH$
Fatty acid alkanolamide	$R\!-\!CONHCH_2CH_2OH$
Fatty acid alkanolamide ethoxylate	$R\!-\!CON{<}^{(CH_2CH_2O)_nH}_{(CH_2CH_2O)_mH}$
Fatty amine ethoxylate	$R\!-\!N{<}^{(CH_2CH_2O)_nH}_{(CH_2CH_2O)_mH}$
Polyalkylene glycol (ethylene oxide/propylene oxide addition products)	$R\!-\!(CH_2CH_2O)_n\!-\!(\overset{\overset{\textstyle CH_3}{\mid}}{CH}\!-\!CHO)_mH$

the use of the nonionic surfactants produced from these raised any problems which had to be regarded as pollution of the environment. Due to the development of petrochemistry on the one hand and under the influence of a shortage of natural raw materials on the other hand, a change occurred in the late sixties. The plantations in Far Eastern countries, in Africa and in America, and the large slaughterhouses in Chicago or Australia, were unable to make available, respectively, sufficient tallow and lard, and coconut oil, palmkernel oil and ground nut oil. The synthetic route, starting from petrochemical raw materials, however, provided adequate quantities of fatty alcohols in the appropriate C-range at economical cost.

At present, the following synthetic routes for detergent alcohols are used:

As early as 1936, Roelen developed the Oxo process [22]. In this process, hydrogen and CO are reacted with olefines, and partially branched alcohols are formed in a two-stage process.

In the Ziegler process, alcohols of a distribution spectrum which corresponds to a Poisson distribution are obtained in a growth reaction from trialkylaluminium to which ethylene is added stepwise to give higher-molecular compounds [23]. These alcohols are exclusively straight-chain and are comparable to natural alcohols.

Moreover, Shell has developed a modified Oxo process in which the proportion of straight-chain alcohols is significantly increased with the aid of phosphorus compounds [24].

With respect to their biodegradability and their toxicity values, the Oxo alcohols or their ethoxylates, as well as the Ziegler alcohol derivatives, meet the legislation in force in the Federal Republic of Germany and in the countries of the EEC.

In the case of alkylphenol polyglycol ether, quantitatively the next most important nonionic surfactant, the alkyl chain used is mostly a tripropylene (nonyl) or dibutene (octyl) chain. The technological properties of these alkylphenol polyglycol ethers are very similar to those having corresponding fatty alcohol derivatives.

The alkylphenol polyglycol ethers have certain advantages for special fabrics, such as say mixed synthetic/wool fabrics.

For a time, these surfactants were regarded as insufficiently degradable. In the meantime, it has been found that a biodegradation of more than 80% can be achieved as a result of adaptation of the bacteria [25].

Various Nonionics

When fatty acids react with ethylene oxide, fatty acid polyglycol esters are obtained which are used as emulsifiers or as special surfactants in cosmetics [26].

Fatty acid amines and fatty amines can also be ethoxylated in the same way. The fatty acid amide polyglycol ethers or fatty amine polyglycol ethers thus obtained are very gentle for the skin and are therefore also used in cosmetics, in the textile industry and in special detergents [27, 28].

The fatty acid monoethanolamides and diethanolamides, which are prepared by the amidation of fatty acids with mono-ethanolamine and diethanolamine, have a superfatting, colloid-stabilising and skin-protective action; in addition, they are good dispersants for the dyeing of textiles and also have an influence on the hydrotropic properties of detergent combinations [29].

Nonionic compounds can also be produced as so-called block copolymers or mixed copolymers with propylene oxide and ethylene oxide. The future use of the products in the washing and cleaning field may be in doubt, since they are not sufficiently biodegradable, unless they contain a straight-chain alkyl group. The products are used as wetting agents and dispersants for lubricants and plasticisers, in the textile, leather, paper and rubber industries, and as heat transfer media and hydraulic fluids, thus in fields of application which are not subject to the Detergent Law, and where no significant quantities are expected to enter the environment.

Amphoterics

In addition to surfactants having a pronounced anionic or cationic character, compounds which have electropositive and electronegative components combined in the same molecule are also important. These compounds behave like anionic surfactants at pH = 8, and cationic surfactants at pH = 4, and they are therefore also called ampholytes or ampho-surfactants.

A further class of amphoteric surfactants, which are related to quaternary ammonium compounds, are the alkylbetaines. These are obtained by reacting an alkyldimethylamine with sodium chloroacetate.

Since they are compatible with the skin, they are used in the cosmetics field [31].

Applications

Detergents for Household and Cleaning Purposes

The production of detergents has continually increased in recent decades. The reasons for this include not only the modern washing processes, but also the increase in the population, the introduction and spread of automatic washing machines, the increase in consumption of textiles and an enhanced awareness of hygiene [32]. All cleaning methods are carried out in an aqueous medium, that is to say the spent wash liquor and the domestic effluents affect water management in the communities and introduce a polution load. This at least partially explains the fact that, when environmental ideas were taken up by the legislature, this sector of industry was affected at a very early stage.

The table which follows shows the composition of high-grade heavy-duty detergents.

Table 3. Laundry detergents: Ingredients of heavy-duty washing powders

Type	Example	Amount (%)
Surfactants	Mixtures of alkylbenzenesulfonates, fatty alcohol ethoxylates or alkylphenol ethoxylates, soap	10 −15
Complexing agents (builders)	Sodium triphosphate Sodium aluminium silicate (zeolite)	30 −40
Bleaching agent	Sodium perborate	20 −30
Corrosion inhibitor	Sodium silicate	3 − 6
Stabilizer	Magnesium silicate	0.2− 2
Anti-redeposition agent	Carboxymethylcellulose	0.5− 2
Auxiliaries	Optical brighteners Perfume Enzymes	0.2− 0.5
Filler	Sodium sulfate	5 −15

The schematic course of a washing process is illustrated in Fig. 6.

The manufacture of detergents causes little pollution of the environment. If possible, purification waters are utilised for the preparation of the detergent suspension, the so-called slurry. Dissolved detergents pass into surface waters where they are rapidly degraded. An unfavourable selection of nonionics (if these contain a high proportion of alcohols which are volatile with water vapour) can cause air pollution, which is described as a pluming effect. Separator filters purify the waste air by removing fine dust particles [57].

In Europe, fabric conditioners are commonly added, which improve the "feel" of the fabric, at the end of the washing process during the last rinsing step. They are also called soft rinses. These are cationic substances which are absorbed on the fibres, without having any washing action. They smooth the fibre surface and impart a pleasant texture to the materials. They also reduce the electrostatic charge on synthetic fibres [33].

An important step in the development of modern washing powders was the replacement, around 1950, of soda as the builder by complexing agents, such as Na diphosphates or, later, Na triphosphates. These complexing agents not only assist the washing process by softening the water, but are essential for the washing process as a whole. Nevertheless, the advantageous technological properties, such as elimination of alkaline earth metal ions, a pronounced washing action on pigment

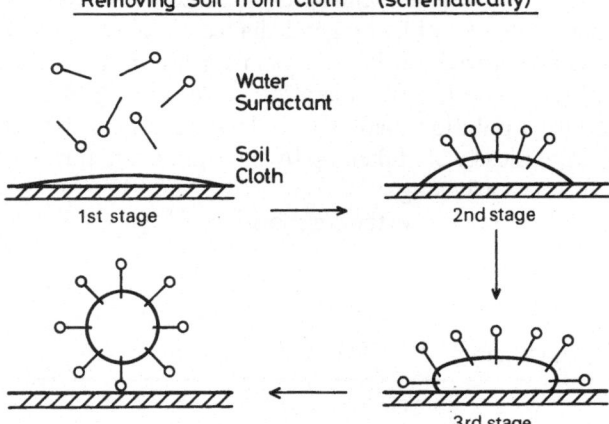

Fig. 6. Schematic presentation of washing process

soils and fat soils, dispersing of the soil particles in the wash liquor, good anti-re-deposition power and favourable corrosion properties, while being toxicologically acceptable for man, were unable to prevent criticism on the phosphates [34]. In stagnant and slow-running surface waters, an excessive growth of algae, due to over-fertilisation (eutrophication) and hence an influence on aquatic organisms was observed. The legislature in the Federal Republic of Germany demanded a solution of the eutrophication problem caused by the detergent phosphates. In the seventies, the chemical industry was asked to develop phosphate substitutes which, on the one hand, have the desirable technological properties of the phosphates and, on the other hand, do not affect the ecosystem. The chemical industry developed new products which would meet the demands. It was found, however, that either the technological properties of the large number of substances developed and tested were inadequate or, if the technological properties were good, the substances were ecologically unsatisfactory. At present, sodium aluminium silicate is used as at least a partial substitute for phosphates [35]. This enables the proportion of phosphate in washing powders to be significantly reduced, and the load on the ecosystem is diminished. As a water-insoluble ion exchanger, this inorganic substance has a high capacity for binding calcium, it is not hygroscopic and it reduces the incrustation on textiles and washing machines. However, this ion exchanger cannot function without certain proportions of water-soluble complexing agents, such as phosphates [36].

A corresponding law, or a statutory order, obliges the detergent industry to reduce the quantity of phosphate in washing powders to about half, in 2 steps. It remains to be seen whether this is sufficient to suppress the eutrophication in our waters [37].

Surfactants for Industrial Purposes

More than one-third of all the surface-active compounds produced are used outside the extensive detergents and cosmetics sector, namely by industry [38].

Manufacturing in the Textile Industry

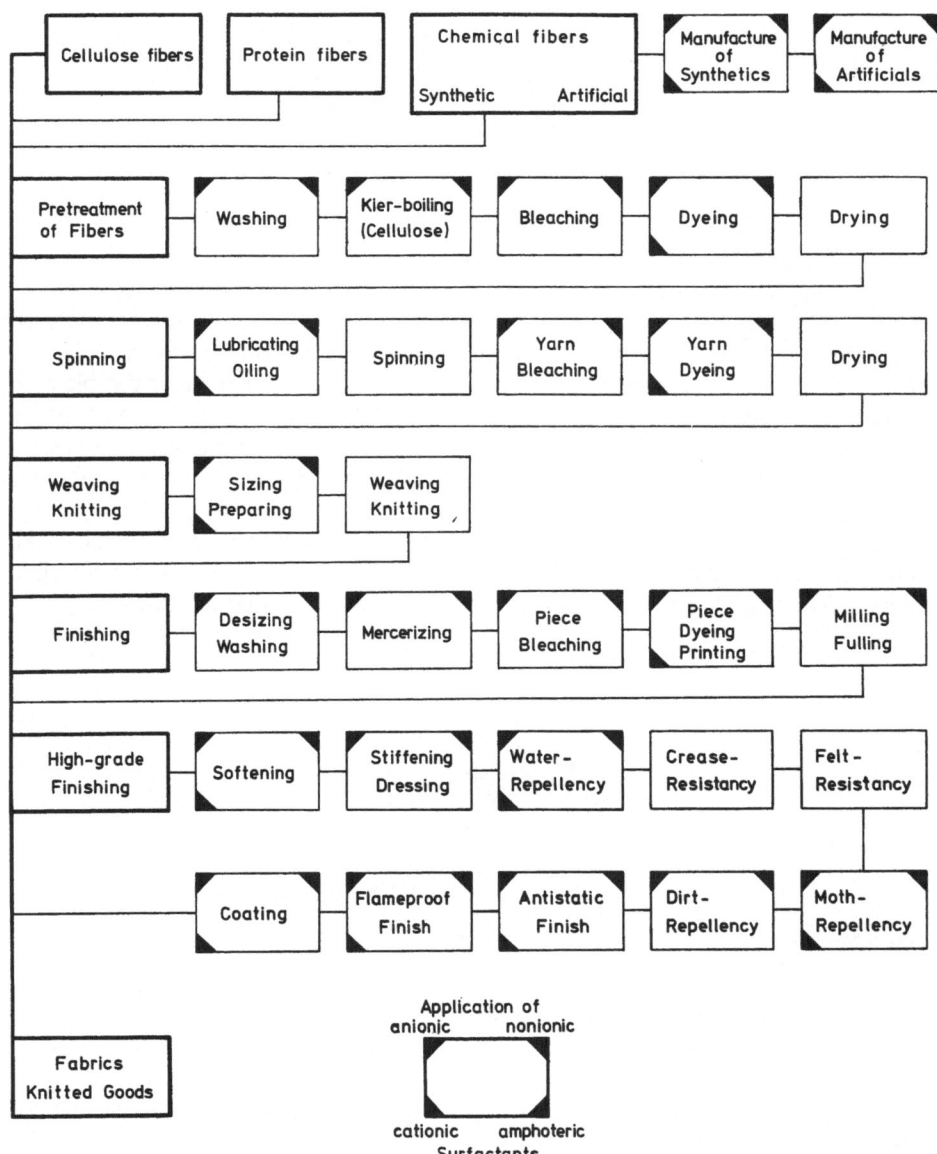

Fig. 7. Working steps in textile manufacturing

Most industries require surfactants not so much for the manufacture of an end product, but rather as indispensable aid in a manufacturing step or in a chemical reaction. This also includes a number of processes which can be regarded as cleaning or washing steps. For textile manufacturing examples of the working steps are diagrammatically shown in Fig. 7.

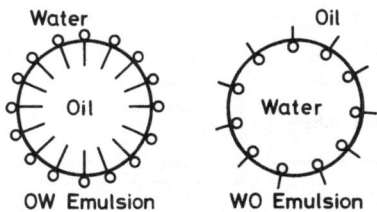

Fig. 8. Schematic representation of oil in water and water in oil type emulsion

Similar problems occur in the leather and fur industries.

No less important, and certainly more problematic with respect to the environment, is the sector where, in addition to cleaning purposes, surfactants are also used for the formation of emulsions for manufacturing processes or chemical reactions. Emulsions are multi-phase systems of liquids which are not miscible with one another [39]. Since the inner, discontinuous phase and the outer, continuous phase in most cases contain water or oil, these types of emulsion are generally called an oil-in-water emulsion (milk type: OW) or water-in-oil emulsion (butter type: WO).

Surfactants as emulsifiers ensure the stability of emulsions. They behave like polar substances which reduce the interfacial tension between the immiscible liquids. Their mode of action can be schematically illustrated as follows [39]:

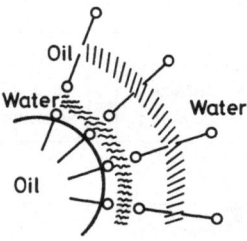

Fig. 9. Schematic representation of emulsifier action

There are a large number of emulsifiers of natural and synthetic origin, which fit into the surfactant classification given. The natural substances used as emulsifiers all fit into this system (See Table 4).

Even today, empirical laboratory experiments are still used for the selection of emulsifiers, although a few fundamental rules exist. Thus, for example, Griffin classified the nonionic emulsifiers numerically between 0 and 20 [40]. A rising num-

Table 4. Natural substances used as emulsifiers

Alginates	Cholesterol	Lignin-sulfonates
Beeswax	Gelatine	Ozocerite
Caseins	Lanolin derivatives	Polysaccharides
Cellulose ethers	Lecithin	Saponins

Table 5. Classification of emulsifiers

HLB range	Application
4– 6	W/O emulsifiers
7– 9	Wetting agents
8–18	O/W emulsifiers
13–15	Detergents
10–18	Solubilisers

ber indicates hydrophilic properties in the form of a hydrophilic-lipophilic balance (HLB value) (Table 5).

Whilst at least the major part of domestic effluents and hence also the detergents and cleaning agents used in the household can be controlled, and worked up in accordance with the latest knowledge by the sewage treatment plants of cities and communities, this is not always the case for industrial effluents. Thus, the biodegradability of emulsifiers or emulsions is a particularly important topic at present. The use of biodegradable emulsifiers cannot by itself solve all the problems. The mineral oil emulsions or solvent emulsions now used in the textile-, metal- or similar industries represent a particular problem. It is necessary to break these emulsions after use by means of chemical or physical methods. One possibility is microbiological decomposition. Furthermore, a separation of the emulsion can be effected by raising or lowering the temperature. Frequently, dilution with the continuous phase or a supply of mechanical energy in the form of centrifuging, stirring or vibration is helpful. The addition of electrolytes, such as common salt, sodium sulfate, iron-III salts and the like, also effects a destruction of the emulsion. When selecting the emulsifiers, those are preferred which can be readily broken. The oil which, after separation, in most cases separates out on the surface can be separated from the residual quantity of water with an oil separator, before the effluent is introduced into the discharge system. During the breaking of the emulsion, a major part of the emulsifiers is retained in the oil which has separated out, and therefore does not pass into the aqueous continuous phase. The separated oil is either worked up or burnt.

The industry is developing so-called de-emulsifiers which, by way of a displacement reaction, exchange the molecules oriented on the interfaces and, due to the reorientation, ensure that the emulsion will be unstable.

Emulsifiers and the emulsions prepared from these are used in very diverse applications. In the foodstuffs industry, fatty acid polyglycol esters or glycerol esters are used for the production of margarine and in the manufacture of cooking fats, milk powders contain emulsifiers in the form of glycerol monostearates, and special lactates and tartaric acid esters of fatty acid glycerides are present in pastries made with yeast dough and in pasta. In most countries, strict food laws limit the use of synthetic emulsifiers [41].

The metal-working and metal-fabricating industries are major users of emulsions for the most diverse metal treatment processes, such as drilling, cutting, milling, sawing and grinding. Large quantities of heat are evolved due to friction and the work of deformation. The heat must be removed in order to reduce wear of the tools and to compensate for the loss in strength. The emulsions used are mostly oil-in-water emulsions, in which lubrication is effected by the oil droplets and the water performs the desired cooling action. The emulsifiers used must be resistant to hardness, dermatologically tolerated and biodegradable, and cause little foaming. Moreover, anti-corrosive properties are desirable. In the non-cutting deformation of metals, emulsions are also employed, in view of the high rolling speed, in order to effect adequate cooling of the roller [47].

Emulsion polymerisation is used at present in the manufacture of plastics dispersions. In this process, the insoluble monomer is emulsified in water. Emulsifiers are also required for the preparation of foamed materials [43].

In the building industry, emulsifiers are widely used for the preparation of bitumen emulsions, and surfactants are used in the production of foamed concretes. Concrete emulsions contain high proportions of paraffins or mineral oils. The boards or plates required in forming are prepared using this emulsion, and they are then less sensitive to dirt and are more readily released [44]. In the chemical industry, in turn, surfactants are frequently added for reasons of reaction kinetics, in order to enlarge the interfaces of the substances which are not miscible with one another [45].

Large quantities of emulsifiers are also used for the manufacture of lacquers and pigment suspensions, and they are used in the photographic industry for the preparation of dye emulsions for colour films [46]. A large field of use is the cosmetic and pharmaceutical industry, even though very strict standards must be applied here. When used in the form of an emulsion, active ingredients and fat components can be used to protect and care for the skin. The fine distribution of the oil component in water, or of the water in oil, enables creams or liquid emulsions to be used to optimum effect. Emulsifiers in cosmetics or pharmaceutical preparations must be subjected to dermatological and toxicological tests [47].

Plant protection and pest control are not feasible without emulsifiers. The preparations are usually marketed as concentrates and, at the point of use, the liquid mixtures or solutions containing the active ingredients and emulsifiers are converted with water into a simple, uncomplicated use form. Emulsions which are guaranteed to have long-lasting stability can be prepared by means of mutually matched mixtures of anionic surfactants and nonionics, together with the insecticidal, herbicidal or fungicidal active ingredients [48].

Economic Significance

The consumption of surfactants in Western Europe is estimated to be more than 1 million tons/year. As shown in Fig. 10, about 60% of this quantity are anionic surfactants (above all LAS), and the remainder is divided into about ¾ of nonionics and ¼ of cationics or amphoteric surfactants.

The economic significance and the application in many fields should be demonstrated once more by reference to the list below, where the specific purposes of the surfactants are also shown alongside the fields of use (Table 6).

By far the major part of the surfactants manufactured is consumed by detergents and cleaning agents. The widely different quantities consumed, depending on

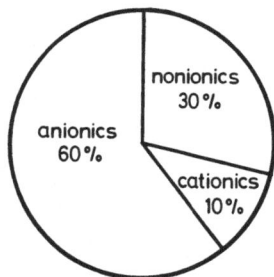

Fig. 10. Consumption of synthetic surfactants in Western Europe (total more than 1,000,000 tons)

Table 6. Application of surfactants

	Typical use
Cleaning agents	
Household	Laundry detergents, dishwashing detergents, household cleaners
Industrial and institutional	Industrial cleaners, floor cleaners, commercial laundry
Personal care	Shampoos, cosmetics
Industrial applications	
Industry	
Textile, leather, fur	Desizing agents, scouring agents, wetting agents, textile softeners
Metal	Surface treatment, rust-proofing, metal-working fluids
Chemical	Emulsifying agents, polymerization agents
Agricultural	Emulsifying and dispersing agents for fertilizers, insecticides and germicides
Pharmaceutical	Emulsifiers, solubilizers, cream stabilizers
Paper, cellulose	Wetting agents, dispersants, de-inking agents
Construction	Air-entraining and foaming agents for cement and concrete
Mining	Flotation agents
Paints, printing	Dispersants
Oil production	Demulsifiers
Food	Emulsifiers for margarine and instant products

the climate, the standard of living and the washing habits, will be seen from a table of the world per capita consumption of synthetic detergents and cleaning agents (excluding soaps) (Table 7).

The large differences between the industrial nations and the developing countries, some of which are at a very low level, are remarkable. Whilst saturation has probably almost been reached in the former, the estimate of the catching-up demand in the countries having a low per capita consumption is high. Since this survey deals only with synthetically produced detergents and excludes soaps, it may be concluded that the catching-up demand for synthetic detergents in the developing countries, where soap is still dominating, will be even larger as a slow levelling-out takes place.

At present, the value of the surfactants marketed in Western Europe is estimated to be about 8,000 million DM/year.

The capacities for the quantitatively most important surfactant, namely LAS, calculated as alkylbenzene, are about 500,000 tons/year in Western Europe and are

Table 7. World per capita consumption of synthetic detergents and cleaning agents in kg

Continent	1960	1966	1968	1970	1974	1976
Western Europe	5.0	7.8	9.3	10.9	14.2	14.4
Eastern Europe	0.5	1.7	2.1	2.7	2.7	4.1
North America	9.8	15.4	16.7	17.7	23.6	25.0
Central America	1.0	1.4	3.2	3.5	4.4	2.8
South America	0.2	0.6	0.6	0.9	2.2	2.8
Oceania	2.9	6.1	7.0	10.0	10.5	11.0
Africa	0.3	0.4	0.5	0.6	0.5	0.7
Asia	0.1	0.3	0.4	0.4	1.1	0.8
World, total	1.4	2.3	2.8	2.9	4.9	4.1

provided by the following large chemical firms:

Shell	United Kingdom
Wibarco	Federal Republic of Germany
Hüls	
Texaco	
Petresa	Spain
Liquichimica	Italy
SIR	

The most important Western European manufacturers of the alcohols required for nonionics, with a capacity of about 350,000 tons/year, are: Shell and ICI in the United Kingdom; Liquichimica in Italy; Henkel and Condea in the Federal Republic of Germany.

The following should also be mentioned for the production of nonylphenols, at about 100,000 tons/year: Hüls in the Federal Republic of Germany; Rhône Poulenc in France; and Liquichimica, Sisas and Montedison in Italy.

In addition to the catching-up demand of the developing countries, a further future increase in the world-wide consumption of surfactants should also be expected. In general, a greater increase is expected in nonionics. This trend is perhaps further reinforced by the necessity of replacing a part of the phosphates in detergents.

Predominantly ecological aspects will have a decisive influence on the continuance and development in the surfactant field, whilst the price, availability and technological properties have, in recent decades, been the main factors in the development of new surfactants and in the optimisation of the types presently known. About 80% of the surfactants used today are based on crude oil. The remainder is provided by natural alcohols and fatty acids. In the long term, major changes can be caused by new aspects with regard to raw material bases.

The survey in Fig. 11 shows the extent to which the surfactants in current use are based, in an integrated raw material sequence, on crude oil and natural gas (Fig. 11).

It remains to be seen whether shortages will occur and whether the price of raw materials depending on crude oil will change to such an extent that those oil seeds which are relatively readily available will gain a greater significance.

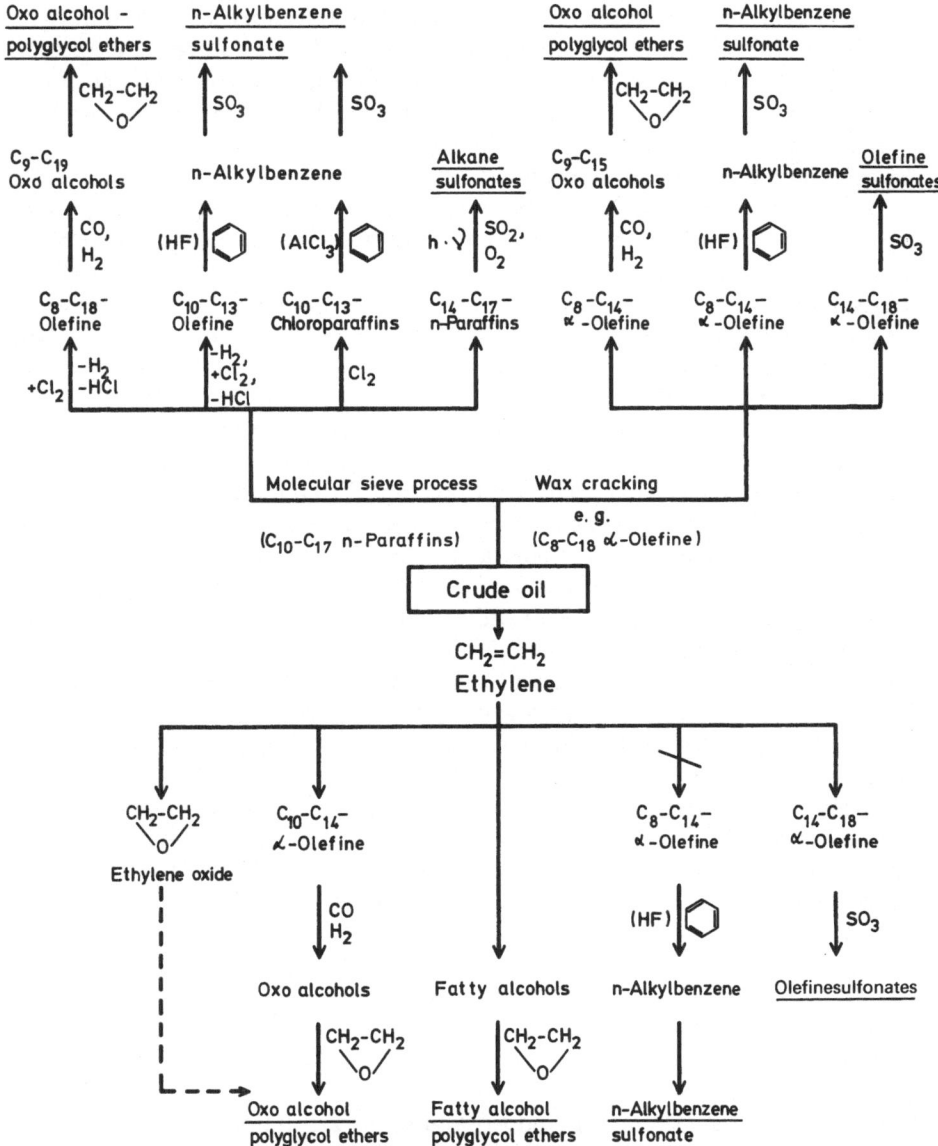

Fig. 11. Raw material bases for detergents

Environment

Introduction

Surfactants are required in a wide range of our daily life for diverse purposes. In many of these fields of application, the surfactants are used in interaction with water. The surfactant develops its beneficial properties in water. After use, however, the surfactant has not been completely consumed, but fairly large quantities are frequently still present in the water. This water containing surfactants then becomes an effluent when it is discharged into the sewers and the treatment plants.

In the treatment plants, the purification of the effluent is carried out mechanically-biologically in most cases. Less often the treatment also includes a third stage, for example via precipitation reactions. Specialised purification processes for particular effluent compositions, for example physical-chemical purification processes, such as have been listed by the Chemical Industries Association [49], are used only in industrial and commercial establishments. With the aid of such special processes, it might be possible in some cases to selectively remove the surfactants used for a special purpose from concentrated effluents of known composition. In most cases, however, the effluent is discharged into a sewer system and passed to central treatment plants.

The surfactants show different behaviour in treatment plants. For example, soaps are precipitated by the hardness in the water and are separated out as insoluble salts in the mechanical primary purification. From there, they pass (with the sludge), into the digestion, which is usually provided in municipal treatment plants, and are readily degraded there under anaerobic conditions. They contribute substantially to the methane produced at this stage [50]. One of the advantages of the modern synthetic surfactants is that within wide limits, they do not form any insoluble salts with the hardness present in the water and are thus not flocculated. A very substantial part therefore passes through the mechanical primary purification, unless they are adsorbed on particles and removed with the latter, a process which is estimated to be very small. The major part of the surfactants thus pass into the biological stage and is subjected to microbial attack.

After passing through the treatment plants, the effluent which has now been purified reaches the receiving bodies of water. Residual products and intermediates which may still be present must not have any detrimental influence on the biological life in the receiving body of water. Rather, they must be amenable to further degradation.

Legislation

In the areas of preparation and industrial use, the legislation relating to the surfactants follows the conventional Water Laws. Production establishments and industrial users are doubtless subject to the normal impositions, for example, those under the Water Resources Law (WHG) [51] in the Federal Republic of Germany. This law, which came into force in 1960 and was amended in 1976, states in § 1 a that water-courses have to be managed in such a way "that any avoidable impairment is prevented". The volume and the treatment of the effluent from production and industrial use are regulated under the stipulations of this law. Accordingly such a degree of purification can be achieved, for example, by measures within the production establishments and by the central treatment plants, for example those having a mechanical and a biological stage, that the stipulations of the Water Laws in the national or international zone are met.

According to the Water Laws, the municipalities would, in fact, be responsible for the same performance within their zone. In conjunction with the detergent legislation, however, the legislature issued a remarkable exceptional provision, prescribing the nature of a commercial material solely on the grounds of water management [52]. Contrary to the rules which are valid otherwise in Water Law, the municipalities were here relieved of the obligation to remove undesirable constituents from the effluent by appropriate purification measures. Rather, for environmental reasons, one demand was made which a product had to fulfill, namely the demand for degradability [53].

Detergent Legislation

Due to the wide distribution of the non-degradable anionic surfactants, in particular tetrapropylenebenzene-sulfonate, in detergents and cleaning agents in the fifties, these products caused a nuisance due to extremely stable foam that was created in many places. This prompted the legislature to act, and this particular field was the first to be regulated by special laws.

A clear account of this development in the Federal Republic of Germany, as given by Dinkloh and Au [53], is presented below, also including an international viewpoint.

In this situation the "Law on detergents in washing and cleaning agents" was passed in the Federal Republic of Germany, and this came into force in 1961 [54]. This law laid down in § 2 that the Federal Government is empowered "to fix by statutory order, in agreement with the Federal Upper House, the demand for degradability of detergents in washing and cleaning agents, and the measurement methods required for this purpose". This statutory order came into force in 1964 [55]. It contains the test method and fixes a percentage of 80% for the required degradation. This demand is raised here only for anionic surfactants which form a chloroform-soluble salt with methylene blue.

Subsequently, OECD began work on the preparation of a recommendation for a generally applicable method for testing the degradability of surfactants. An international group of experts worked out appropriate proposals which were published by OECD in 1971 [56] and were later adopted by the EEC [57].

In an EEC guideline of 22nd November 1973 for harmonising the legal regulations of the member countries with regard to detergents [57], rules common to the member countries are given. According to Article 2, the member countries prohibit the marketing and use of detergents if the average biodegradability of the surface-active substances of any category contained therein is less than 90%. Approval may be refused only if the analysis shows that the biodegradability is less than 80%; in other words, an absolute lower limit without tolerances has been fixed in this case.

The guideline on the methods for checking the biodegradability of anionic surface-active substances [58] prescribes the methods to be used.

The permitted tests are a method approved in France [58], the German method [58] and the OECD method [56]. In cases of doubt, the OECD Confirmatory Test is used as a reference method and, accordingly, is described in detail [58].

An amended and extended "Law on the compatibility of detergents and cleaning agents with the environment (Detergent Law)" has come into force in the Federal Republic of Germany on 1.8.1975 [59]. According to the leading sentence (§ 1), detergents and cleaning agents may "be marketed only if, after their use, a detrimental change in the quality of surface waters, in particular with regard to the drinking water supply, and an impairment of the operation of treatment plants are not to be expected".

§ 2 gives the following definition: "Detergents and cleaning agents within the meaning of this Law are products which, as a mixture of surface-active substances and phosphates, perborates or substitutes therefor and other complementary constituents, or by means of one of these substances, have, in interaction with water, a cleaning effect or are intended to be used for cleaning and can pass into watercourses after use". With this definition, the German regulation goes beyond the earlier legal stipulations, which related only to the surfactant components, and comprises all the constituents of a detergent and cleaning agent, independently of the surfactant content. Moreover, the law empowers the Federal Government to fix, by statutory order, the required degradability, or the removal by other means, of surface-active or other organic substances contained in detergents and cleaning agents, and to fix the measurement methods required for this purpose. Moreover, the phosphate content in products of this type may be regulated by statutory order. Other ingredients capable of having a detrimental effect may likewise be subjected to regulations. Moreover, the statement of data on the package, the statement of recommended dosages corresponding to the hardness of the water used in practice, and the submission of basic recipes to the Federal Environment Office are regulated. Supervison is the duty of the authorities competent under state law.

In follows that the law not only covers detergents and cleaning agents in the domestic sector, but also includes the industrial sector, for example the textile industry. In this respect also, the new law goes beyond the 1961 regulations.

Whilst it was common to all the earlier laws and regulations that they made stipulations only with respect to the degradability of anionic surfactants, the OECD investigations and tests in 1976 for the first time included the nonionic surfactants [60]. The German legislation likewise has also included the nonionic surfactants and laid this down in a corresponding statutory order [61]. In the meantime, the EEC authorities have taken up these problems and also worked out guidelines for this field, which at present are available in draft. Publication is to be expected in the near future. Trade restraints in the EEC territory are thus prevented by unified regulations, with a single arbitration method for cases of doubt, whilst at the same time national methods are mutually acknowledged.

Water Resources Law and Related EEC Regulations

The laws discussed so far, namely the Detergent Law and the corresponding EEC guidelines, make demands on the products, that is to say they are economic laws. A different situation, however, applies in the effluent and water-courses sector, where requirements relating to the quality of water-courses or the purification of effluents, before they are discharged into receiving bodies of water, are laid down by the national laws and the international guidelines.

In the Federal Republic of Germany, the Water Resources Law (WHG) [62] was passed in 1957. As an enabling law, the execution of which is within the competence of the states, this law, together with the state water laws and the relevant regulations, regulates the entire water industry in the Federal territory. The law stipulates that any utilisation must be carried out in such a way that an adverse biological, chemical or physical change in the water-course cannot occur. Discharges are compulsorily subject to consent or approval. Limitations in water by-laws, which comprise measures to counter the adverse effects of a discharge, for example by adequate treatment of the effluents before discharge, are laid down by the competent authorities, taking into account the relevant statutory regulations, for the particular case.

In the EEC territory, various guidelines have been issued which deal with the purity of water-courses, for example guidelines on the discharge of effluents or on the suitability of water-courses for treatment as drinking water.

In contrast to the Detergent Laws, the general water legislation does not cover the anionic and nonionic surfactants specifically, but it covers surfactants in the same way as any other components present in effluent; however, this is not done via product-specific analytical methods, but within the ambit of the particular requisite parameters, or at least via general parameters such as, for example COD [chemical oxygen demand] or DOC [dissolved organic carbon]. However, this is then no longer specific to surfactants, so that a further description is unnecessary.

Analytical Methods

To properly assess the role of surfactants in the environment, the nature and concentration of these compounds in effluents and surface waters must be known. This requires reliable product-specific analytical methods. Summarising descriptions were recently published by Wickbold [63] and Kunkel [64]. Analytical methods are necessary to verify environmental levels and to follow disappearance of parent compounds in biodegradation tests. The definition of the OECD report [60] reads as follows:

"In the present context ‚biodegradation' signifies the breakdown of an organic compounds by micro-organisms, under the conditions of the test methods, resulting in the loss of certain initial characteristics environmentally undesirable.

In the case of anionic surfactants the course of biodegradation is followed by an analytical procedure based on the formation of chloroform-soluble salts with methylene blue, and expressed in terms of a known tetrapropylenebenzene.

In the case of nonionic surfactants the course of biodegradation is followed by an analytical procedure (method of Wickbold) based on the formation of a precipitate with barium tetraiodobismuthate. In its present form the procedure refers specifically to water-soluble ethoxylates and propoxylates. This group comprises the bulk of nonionic surfactants presently in use."

Corresponding to this definition, the analytical methods are designed to cover intact anionic and nonionic surfactants.

In some cases, in particular if the concentrations are very low, for example in surface waters, it is advisable to concentrate the surfactants before analysis. For this purpose, the water sample is covered in a cylindrical vessel with a layer of ethyl acetate and a stream of nitrogen saturated with ethyl acetate is then carefully blown through. The surfactants are transported on the surface of the nitrogen bubbles into the organic phase and are concentrated there. A further advantage of this step is that other substances which are present in the water and which interfere with the surfactant analysis are separated. These are not blown out and concentrated in the ethyl acetate phase, but they remain in the aqueous phase, which is discarded. The organic phase is separated, the solvent is evaporated and the residue is taken up in water and analysed for surfactants [65].

Longwell-Maniece Method (Methylene Blue Method) *for Anionic Surfactants* [58, 60, 66]

In this method, the water-insoluble methylene blue salt of the anionic surfactants is transferred into chloroform by shaking. To eliminate any interference by other substances which react with methylene blue such as proteins and humic acids, the chloroform extract is shaken with an acid methylene blue solution. The blue chloroform extract, made up to the correct volume, is subjected to photometry and the result is evaluated by reference to a calibration curve. An advantage of this method is that it can be automated. Soaps do not react with methylene blue.

Wickbold Method (Barium Bismuth Iodide Method) *for Nonionic Surfactants* [66]

Nonionic surfactants in a water sample or in the residue taken up with water are, before blowing out, precipitated as a surfactant/$Ba(BiI_4)_2$ complex by means of the Dragendorff reagent. After filtration and washing with glacial acetic acid, the precipitate is dissolved in ammonium tartrate solution and the bismuth is then titrated potentiometrically with a solution of pyrrolidine dithiocarbamate [62, 66].

Determination of Cationic Surfactants with Disulfine Blue

According to Kunkel [67], the cationic surfactants are, by blowing out, concentrated in ethyl acetate and largely separated from interfering non-surfactant substances. Additions of an anionic surfactant and a neutral salt to the water sample ensure a complete transfer into the organic phase. After separation of the layers and evaporation of the solvent, the anionic surfactants are separated off by means of ion exchangers and the cationics are determined photometrically by a modified disulfine blue method. To avoid troublesome turbidity in this method, the dye complex formed is removed from the $CHCl_3$ solution and redissolved in methanol.

The Michelsen method [68] proceeds in a similar manner, but the nonionic surfactants are determined by means of thin-layer chromatography.

Waters and Kupfer [69] evaporate the water sample for concentration, take up the residue in a solvent and, after ion exchange, proceed in the same way as in [68].

The method covers cationic surface-active substances of the type of quaternary nitrogen compounds, such as $(R_4N)^-$ and the like, having molecular weights from about 250 to 600.

The methods are suitable for the investigation of water and effluent and also for the evaluation of degradation tests. The cationic surfactant concentrations to be expected in these cases are normally in the ppb range ($= \mu g/l$).

Environmental Behaviour

Biological Degradation Tests

Test Methods

In view of the situation in treatment plants and receiving bodies of water, aerobic degradation is the most important biological process to be investigated.

When selecting a suitable test method, the predominant factor must be that it is applicable in practice. This means that already known and proven microbiological laboratory methods cannot be adopted without further modification. Thus, for example, the use of selective pure cultures, or of mixed cultures preadapted to the substrate in the case of specific problems, can give interesting results, without solving any problems closely related to the natural conditions in effluent treatment and surface waters.

In such cases metabolic competition of the surfactants, which in general is unfavourable under natural conditions, may have been neglected or test methods may be inadequately defined and cannot be carried out in a routine manner.

Considering these aspects, two test methods have been developed for the determination of the degradability of surfactants which simulate the conditions in effluent purification (continuous activated sludge process) or in the receiving body of water (static screening test). Because of their relative simplicity they can be carried out in large series as a matter of routine.

Summarising descriptions have been published by Bock and Schöberl [70, 71].

OECD Confirmatory Test (Activated Sludge Test)

This test method comprises a continuously operating activated sludge test which is recommended by OECD as a "Confirmatory Test" [60] for anionic surfactants and has been included in the relevant EEC guideline [58] in the form of a reference test.

Figure 12 shows a diagrammatic illustration of the prescribed acitvated-sludge test unit. It operates as follows:

With the aid of a liquids pump B, 1 l of synthetic effluent which contains mineral salts, peptone, meat extract and 20 mg/l of the anionic surfactants to be tested or 10 mg of the nonionic surfactants to be tested is delivered per hour from a supply vessel A into the aerated reaction vessel C which has a useful capacity of 3 l.

Due to a spontaneous infection during the so-called breaking-in period, an activated sludge (bacteria, possibly fungi, protozoa, flagellates, nematodes and the like) is formed in the aeration vessel C and this passes with the effluent into the settling vessel D. The treated water passes from this vessel into the receiver F, whilst the sedimented activated sludge is recycled with a mammoth pump E ("air-lift" method) into the reaction vessel C. The air rate introduced into the reaction

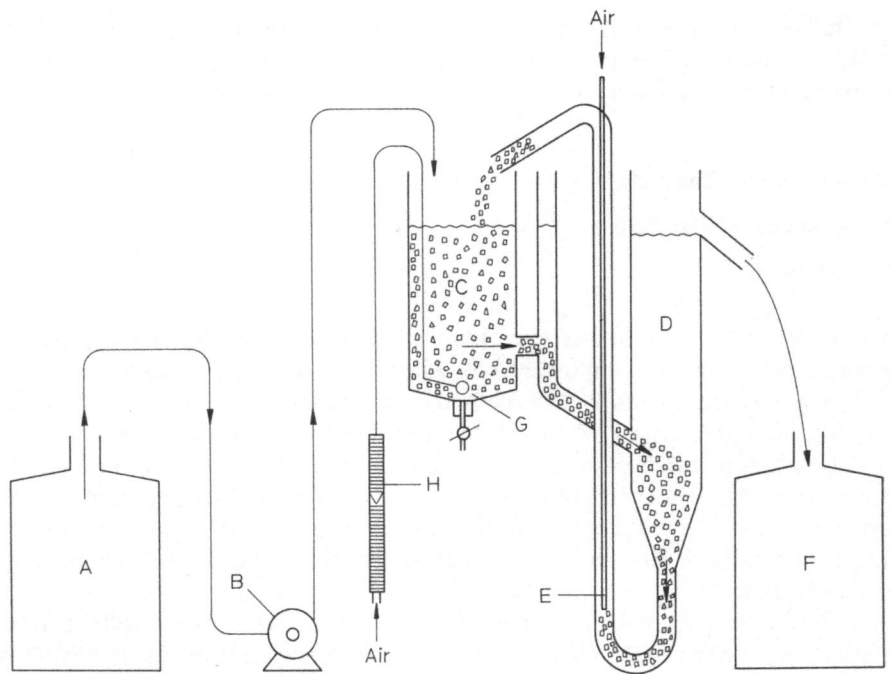

Fig. 12. Diagrammatic illustration of the OECD Confirmatory Test [60]
A Supply vessel, B Metering device, C Aeration vessel, D Settling vessel, E Airlift pump, F Receiver, G Frit, H Air flowmeter

vessel is such (measuring instrument H) that the oxygen content in the vessel C does not fall below 2 mg/l. At the same time, the content of the vessel is maintained in suspension (stirring effect) by the aeration (frit). The residence time in the basin C is 3 h, analogously to the conditions obtaining in practice in treatment plants (the feed rate to the test apparatus is 24 l per day).

The corresponding surfactant contents are obtained by chemical analysis of the inflow and outflow, and the degraded proportion in per cent is calculated from these figures.

Figure 13 shows typical degradation curves for surfactants of high and low biodegradability.

OECD Screening Test

As already mentioned, the screening test described below is intended to give a close simulation of the situation in surface waters. This test has additional importance as screening test method. Since the activated-sludge test cannot be applied reactively in every case, because it requires a relatively large effort, it is necessary to use less complicated laboratory tests for obtaining preliminary information on the degradability. The OECD Screening Test which, in the OECD test methods [60] for the investigation of the degradability of anionic surfactants, is carried out be-

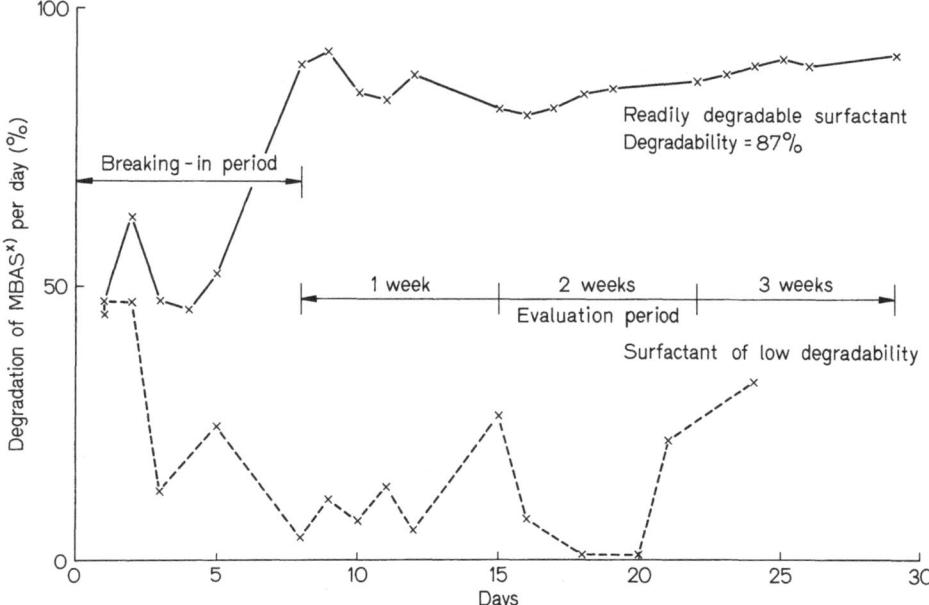

Fig. 13. Degradation curves for surfactants of high and low degradability in the OECD Confirmatory Test [54]
×MBAS = methylene blue-active substance

fore the Confirmatory Tests, is designed as a so-called "end-point test": starting with an initial concentration, in this case 5 mg/l of substance to be tested, the degradation curve is analytically determined, until stabilisation occurs, in a precisely defined arrangement, and the end value is calculated as a percentage surfactant loss.

The investigations are carried out in shaking flasks, with the surfactant to be tested being the sole source of carbon and energy in a mineral salt solution.

In order to ensure that reproducible results are obtained in different laboratories, only a small amount of a polyvalent bacteria suspension which has not been preadapted (for example discharge from an aerobic treatment plant, in most cases a municipal treatment plant) is used for inocculation. To check the system, a biologically "hard" standard (tetrapropylenebenzenesulfonate) and a biologically "soft" standard (MARLON A®, a linear alkylbenzenesulfonate) of known degradation behaviour are tested in parallel with each test series. The maximum duration of a test is 19 years.

According to the OECD instructions, the results of the Screening Test need to be supported by the Confirmatory Test only if the results are below or just within the range of the statutorily required degradability of 80%, or if the results appear to be doubtful.

Biodegradation

In the statutorily prescribed methods [58, 59, 61], the degradability is determined by analytical methods which are specific for the substance, i.e. the methylene blue

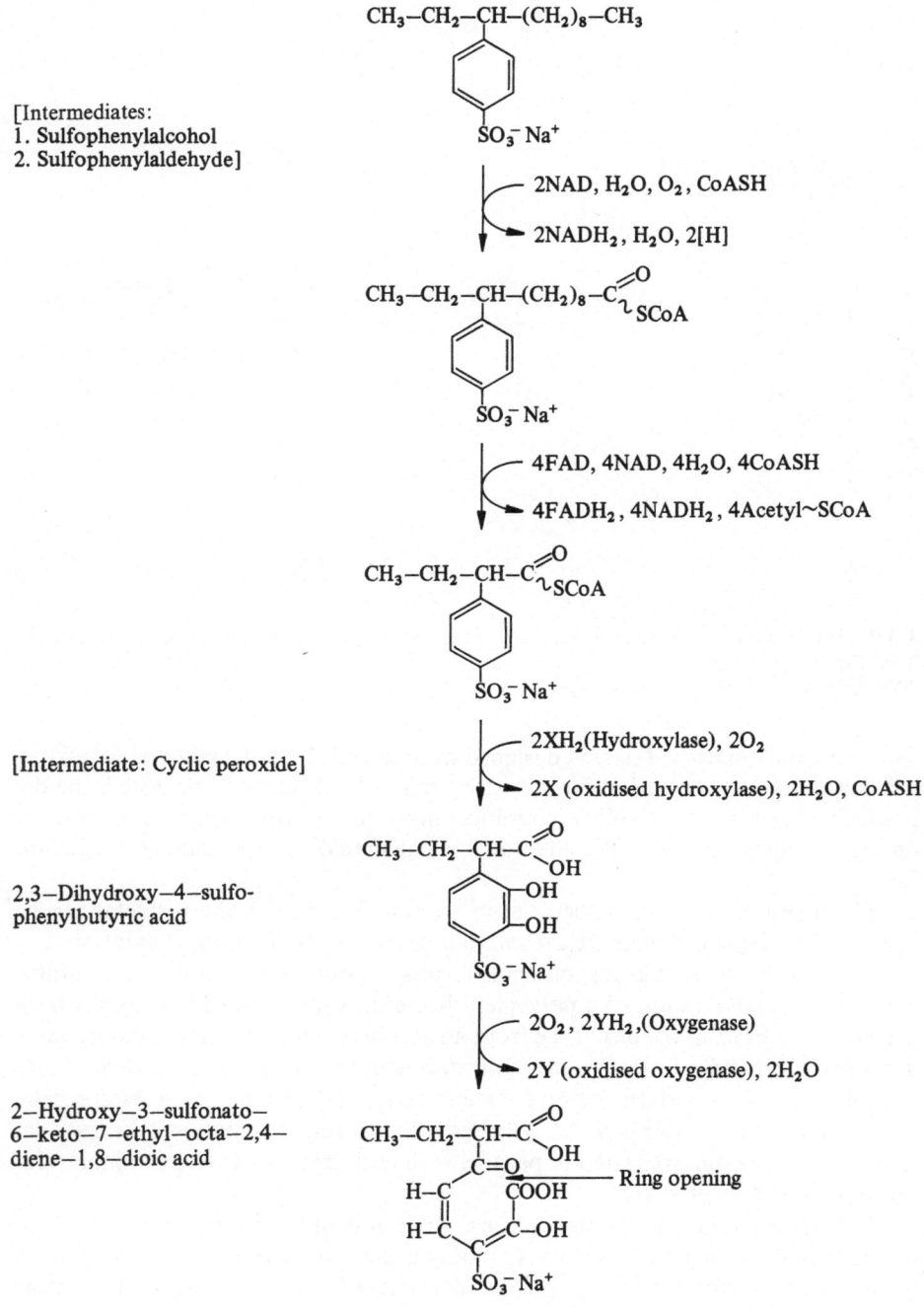

[Intermediates:
1. Sulfophenylalcohol
2. Sulfophenylaldehyde]

[Intermediate: Cyclic peroxide]

2,3–Dihydroxy–4–sulfo-
phenylbutyric acid

2–Hydroxy–3–sulfonato–
6–keto–7–ethyl–octa–2,4–
diene–1,8–dioic acid

Fig. 14. Diagram of the biochemical oxidation of alkylbenzenesulfonate (including ring opening). (From [72])

method for anionic surfactants, and the determination of the bismuth-active substance for nonionic surfactants.

No generally prescribed analytical method for cationic surfactants has as yet been recommended.

It may be assumed that all the anionic and nonionic surfactants used in detergents and cleaning agents exceed the statutorily required minimum degradation of 80%. This was and is checked in numerous laboratories, in some cases as a matter of routine. Investigations in large-scale treatment plants confirm these results (for example [73]).

In addition to the knowledge gained by analytical methods specific to a substance, extensive investigations into the route of the microbial degradation and remineralisation are available, and these have been reported in numerous publications.

As early as 1964, R. Krüger was able to state in summary [74], as "new results in the alkylbenzenesulfonate field", that the very extensive remineralisation of the most important anionic surfactant, namely linear alkylbenzenesulfonate, takes place by opening of the aromatic ring in the course of biological effluent treatment. It is known that this degradation, apart from the first oxidation step, does not require any specific metabolic mechanism. Only the first attack on the aliphatic chain or on the aromatic nucleus requires special enzyme systems. Once the molecule has been "broken open" by the introduction of oxygen, the surfactant undergoes biochemical degradation similar to most organic substrates. Reviews of this topic have been published by Swisher [75] a few years ago and by Schöberl and Bock [72] more recently.

Accordingly, the microbial degradation proceeds via the following steps (Figures 14 and 15).

Investigations by Steber [76] using linear alkylbenzenesulfonate (MARLON A®) labelled with C^{14} in the ring confirm the degradation of the aromatic ring and the total degradation of $>80\%$.

The degrees of degradation of linear C_{10-13}-alkylbenzenesulfonate determined by very diverse analytical methods, but using the same degradation method (OECD Confirmatory Test) are shown in Table 8.

A great deal of work has also been carried out on the biodegradation of nonionic surfactants, in this case the ethoxylates used in detergents and cleaning agents. However, the metabolic mechanism has not been investigated to the same extent as with linear alkylbenzenesulfonate. The microbial attack takes place on the ethylene oxide chain and also on the alkyl group of the molecules.

Summarising the microbial degradability of the surfactants which are important for use in detergents and cleaning agents, according to the investigation by the OECD Confirmatory Test, using substance-specific analytical methods (methylene blue activity, bismuth activity, dissolved organic carbon, ^{14}C methods and degradation of metabolites), the following picture is obtained [72] (Table 9):

Microbial degradation of surfactants, as it proceeds in biological treatment plants, and also in surface waters under the influence of microorganisms is very extensive. The products are thus "biodegradable" according to the statutory regulations (analysis for MBAS and BiAS) which is almost complete in some cases. Intermediates which occur are amenable to a further biodegradation which proceeds

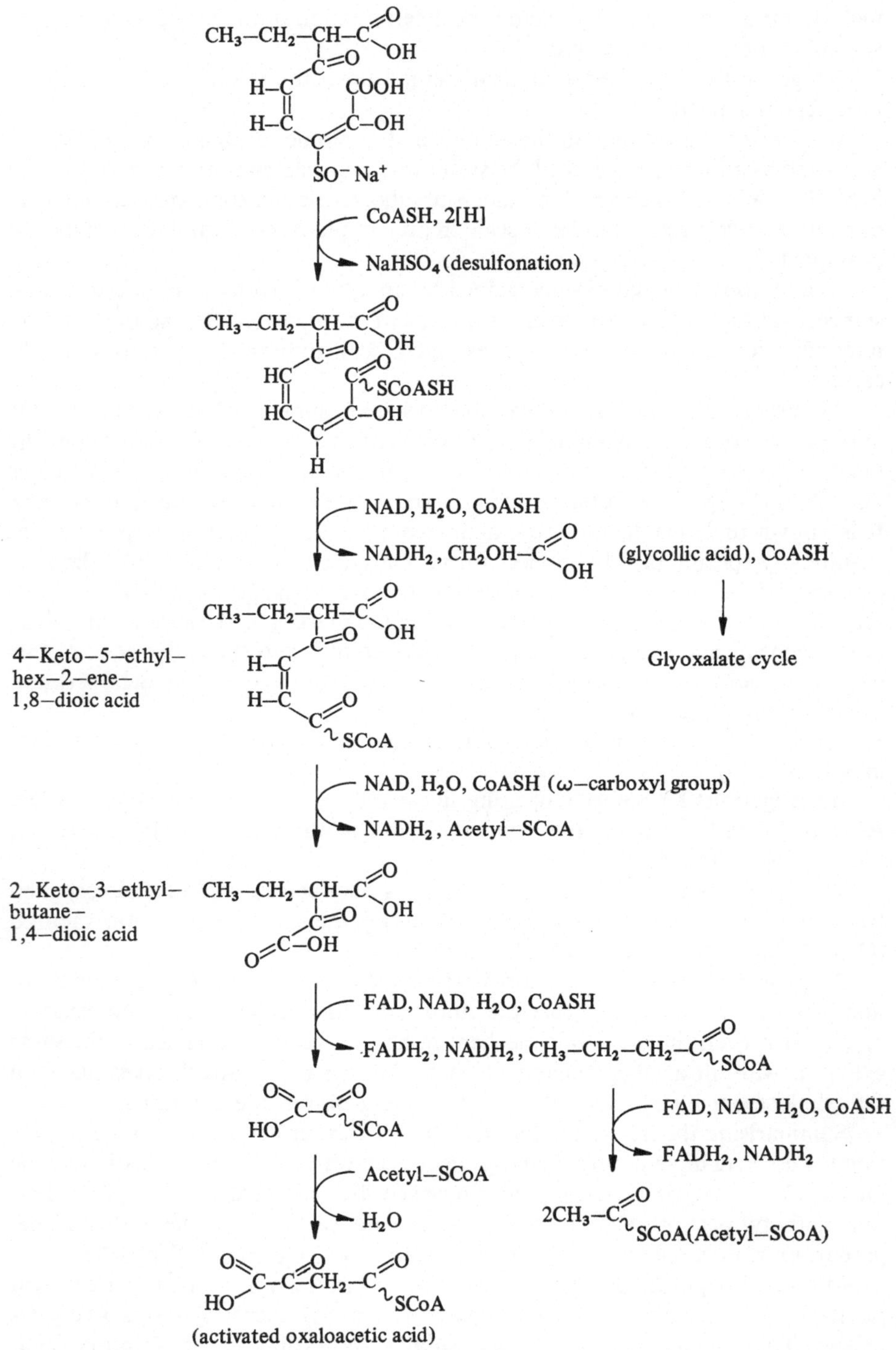

$$CH_3-CH_2-CH-C\underset{OH}{\overset{O}{\diagup}}$$
$$C=O$$
$$H-C \qquad COOH$$
$$H-C \qquad C-OH$$
$$SO^- Na^+$$

CoASH, 2[H]

NaHSO$_4$ (desulfonation)

$$CH_3-CH_2-CH-C\underset{OH}{\overset{O}{\diagup}}$$
$$C=O$$
$$HC \qquad C\overset{O}{\diagdown}SCoASH$$
$$HC \qquad C-OH$$
$$H$$

NAD, H$_2$O, CoASH

NADH$_2$, CH$_2$OH$-$C$\overset{O}{\underset{OH}{\diagup}}$ (glycollic acid), CoASH

Glyoxalate cycle

4$-$Keto$-$5$-$ethyl$-$
hex$-$2$-$ene$-$
1,8$-$dioic acid

$$CH_3-CH_2-CH-C\underset{OH}{\overset{O}{\diagup}}$$
$$C=O$$
$$H-C$$
$$H-C \quad C\overset{O}{\diagdown}$$
$$SCoA$$

NAD, H$_2$O, CoASH ($\omega-$carboxyl group)

NADH$_2$, Acetyl$-$SCoA

2$-$Keto$-$3$-$ethyl$-$
butane$-$
1,4$-$dioic acid

$$CH_3-CH_2-CH-C\overset{O}{\underset{OH}{\diagup}}$$
$$C=O$$
$$O=C-OH$$

FAD, NAD, H$_2$O, CoASH

FADH$_2$, NADH$_2$, CH$_3-$CH$_2-$CH$_2-$C$\overset{O}{\diagdown}$SCoA

$$O=C-C\overset{O}{\diagdown}$$
$$HO \qquad SCoA$$

FAD, NAD, H$_2$O, CoASH

FADH$_2$, NADH$_2$

Acetyl$-$SCoA

H$_2$O

2CH$_3-$C$\overset{O}{\diagdown}$SCoA(Acetyl$-$SCoA)

$$O=C-C\overset{O}{=}CH_2-C\overset{O}{\diagdown}$$
$$HO \qquad\qquad\qquad SCoA$$

(activated oxaloacetic acid)

Fig. 15. Diagram of the biochemical oxidation of alkylbenzenesulfonate (after ring opening). (From [72])

Table 8. Biodegradability of C_{10-13}-alkylbenzenesulfonate (MARLON A®). (From [72])

Analytical method	Degree of degradation (OECD Confirmatory Test)
1. MBAS analysis	∅ 96
2. DOC analysis	∅ 73 (coupled units test)
3. COD analysis	∅ 66 (coupled units test)
4. UV analysis	69 (3 h residence time) and 85 (6 h residence time)
5. ^{14}C analysis of metabolite fractions (^{14}C = aromatic C)	80–85

Table 9. Biodegradability in the OECD Confirmatory Test. (From [72])

Compound	MBAS	DOC	^{14}C	Metabolites
I. Anionic surfactants				
1. Alkanesulfonate (C_{14}–C_{12}–AS)	98%	85–90%		
2. Alkylbenzenesulfonate (linear C_{10}–C_{13})	96%	70% (65% COD)	80%	>80% (UV analysis)

Compound	BiAS	DOC	^{14}C	Metabolites
II. Nonionic surfactants				
1. Fatty alcohol ethoxylate (linear C_{12} plus 12 mols of EO)	>90%	62–76%		75% (mass balance)
2. Oxo alcohol ethoxylate (C_{13}–C_{14})	95%	60%		80% (mass balance)
3. i-Nonylphenol ethoxylate (9 mols of EO)	>90%	76%		77% (mass balance)

very rapidly under the conditions of biological effluent treatment. No accumulation or adverse effects, for example in the receiving body of water, have so far become known.

The results of the laboratory experiments were confirmed in various, sometimes very extensive investigations in treatment plants and water-courses. Earlier reviews were published by Heinz and Fischer [77] in 1962, and by Husmann, Malz and Jendreyko [78] in 1963. Full-scale tests on percolating filter plants confirmed the values found in the laboratory for the degradation of linear alkylbenzenesulfonate [79]. During these tests, it was also observed that a part of the surfactant is already degraded, before it reaches the treatment plant, while the effluent runs along the sewer system [80].

Equally favourable results have been obtained for the degradation of nonionic surfactants in treatment plants, even though fewer investigations have been carried out [70].

May and Neufahrt [81] have reported on the behaviour of the cationic surfactant distearyldimethylammonium chloride (DSDAC) in sludge-activating units. In the OECD Confirmatory Test, they obtained elimination rates of between 91% and 93% at a feed rate of 2.78 mg of DSDAC/l or 1.75 mg of DSDAC/l, with an addition of anionic surfactants or anionic and nonionic surfactants. There is no adverse effect on the anaerobic digestion.

Huber [82] concludes in his review that no adverse effects on effluent treatment and water-courses are known, even though further elucidation of the biochemical degradation route and of the degradation behaviour in treatment plants under conditions applying in practice is still necessary.

Concentrations in Natural Waters

Recently, W. K. Fischer gave a review of "The evolution of surfactant concentrations in German natural waters 1960–1980" [83], following earlier investigations [84] carried out over many years. In addition to the concentration of anionic surfactants obtained by the methylene blue analysis (MBAS), it has also been possible (from 1972) to monitor the values for nonionic surfactants present in natural waters, with the bismuth analysis (BiAS). Fischer found that in the Rhine and in its tributaries, an overall steep decrease of concentrations took place, down to very small residual values (about 0.02–0.1 mg of MBAS/l) as a yearly average, which were only rarely exceeded. He found a significant correlation to the extension of biological treatment plants, which had been carried out in this period.

Fischer points out that the values obtained by the MBAS analysis in the investigation of natural waters are too high in this trace range. The special analysis technique according to Waters [85] shows that the contents of true anionic surfactants, for example in the Rhine, are in the range from 15 to 35% of the MBAS value, which is extremely low.

According to Fischer, 0.02–0.1 mg/l of nonionic surfactants (BiAS) are present; higher concentrations are found only rarely.

By approximate calculation the consumption of surfactants in the catchment area of the Rhine (about 40 million inhabitants) were compared with the anionic and nonionic surfactants actually found [83]. Of the substance originally used, only 1.2% of the anionic surfactants and 2.7% of the nonionic surfactants were still present in the Rhine at Emmerich, upstream of the Dutch border, according to the mean values of the concentrations in 1979. Considering relevant loads, the figures indicate 0.68% of the anionic surfactants consumed and 1.59% of the nonionic surfactants consumed.

Huber [82] gives concentrations of cationic surfactants of 6–20 ppb in the Main near Frankfurt, and 2–30 ppb (= 0.01–0.03 mg/l) in the region of the confluence of the Main and the Rhine.

Toxicology

Lower Organisms

As outlined in the previous sections very extensive microbial attack takes place in the course of the biological effluent treatment. Microorganisms which degrade surfactants are present already in the sewer system before the treatment plant as well as in the latter, and also later in the natural waters. An adverse effect of the cur-

Table 10. Acute lethal concentration of linear alkylaryl-
sulfonates homologues in golden orfes

LAS Homologues	LC_{50} (mg/l) Leuciscus idus melanotus
C_{10}	16.6
C_{11}	6.5
C_{12}	2.6
C_{13}	0.57
C_{14}	0.26
C_{16}	0.68

rently used products in the observed concentrations on the microorganisms active in these ecosystems is not detectable.

Fish are regarded as significant indicator organisms for the quality of natural waters. An investigation of the effect on fish is therefore *inter alia* an important pointer in the assessment of the products.

A first systematic investigation on the effect of increasing concentrations of linear alkylbenzenesulfonates on fish by Hirsch [13, 86] shows that their acute lethal concentration (LC_{50}) depends on the length of the alkyl chain. The LC_{50} for golden orfes in the test with LAS homologues of different alkyl chain lengths is indicated in Table 10.

Additionally, the position of the benzene nucleus on the alkyl chain also influences the LC_{50} [86]. Isomers with a terminal arrangement of the benzene nucleus are more toxic than individual compounds with the nucleus in a central position. The observed effect of surfactants of this type on fish is a swelling of the gill epithelium, which, in turn, is due to lowering of the surface tension [87].

The relationships between the use in practice, the degradation behaviour and the effect on fish of linear alkylbenzenesulfonates are shown in Fig. 4. As a result, an alkyl chain length of C_{10-13} was selected for the alkylbenzenesulfonate mainly used today [13, 74, 86], as described earlier in this chapter.

During microbial degradation, the configuration of the alkylbenzenesulfonate molecules is changed, and the surface activity disappears at an early stage. The intermediates and end products of biodegradation are markedly less harmful to fish; this was already pointed out, as a result of investigations in practice, by Herbert and collaborators and by Niemitz and Pestlin [88]. Later, Schöberl and Kunkel [89] were able to provide experimental proof by means of the isolated metabolites.

These data show that deaths of fish in natural waters under the conditions prevailing today, have not been observed in countries such as the Federal Republic of Germany [90].

The acute toxicity (LC_{50}) of the nonionic surfactants most commonly used in detergents and cleaning agents (fatty alcohol ethoxylates with 10–40 mols of ethylene oxide or alkylaryl polyglycol ethers with about 10 mols of ethylene oxide) is in the range of 3–7 mg/l of active substance in a static test with golden orfes. Similarly to anionic surfactants, the toxic action of nonionic ethoxylates on fish decreases with the chain length, as pointed out by Alabaster [91].

For the cationic surfactant distearyldimethylammonium chloride, an acute fish toxicity of about 2 mg/l is indicated by a 48 hour laboratory test; under conditions

Table 11. Mammalian toxicity of surfactants (From [92])

Substance	Acute toxicity LD_{50} [mg/kg] Animal species	Chronic toxicity dosage Administration period Findings
Alkylbenzenesulfonate (Na salt) C_nH_{2n+1} (linear) ⟨◯⟩—SO_3Na	650 [33] 1.260 [34] Rats	Rats 5.000 ppm i.f.[a] 2 years No findings [35]
Alkylbenzenesulfonate (Na salt) (= tetrapropylenebenzenesulfonate) C_nH_{2n+1} (branched) ⟨◯⟩—SO_3Na	1.220 Rats [34]	Rats 5.000 ppm i.f.[a] 2 years No findings [36]
Alkanesulfonate $C_nH_{2n+1}(SO_3Na)$	3.000 [37] Rats	Rats 0.1 LD_{50}/day 45 days No findings [37]
Olefinesulfonates C_2H_{2n+1}—CH=CH—CH_2—SO_3Na C_2H_{2n+1}—CH—CH_2—CH_2—SO_3Na　　　　　|　　　　　OH	3.600 [38] Rats	—
Alkyl. sulfate C_nH_{2n+1}—OSO_3Na $N=$　8 　　　10 　　　12 　　　14 　　　16 　　　18	3.200 [39] 1.950 2.640 3.500 3.000 3.000 Rats	Rats 1.000 ppm i.f.[a] 90 days No findings [40]
Dodecyl ether-sulfate (3 EO) $C_{12}H_{25}$—$(OCH_2$—$CH_2)_2OSO_3Na$	1.820 [41] Rats	Rats 0.1; 0.5% concentration i.f.[a] 2 years No findings [41]
Lauryl polyglycol ether (7 EO) $C_{12}H_{25}$—$(OCH_2$—$CH_2)$—OH	4.150 [42] Rats	Rats 1.17% maximum i.f.[a] 4 weeks No findings [42]
Nonylphenol polyglycol ether (9 EO) C_9H_{19}—⟨◯⟩—$(OCH_2$—$CH_2)n$—OH	2.600 [43] Rats	Rats 1.14% kg/day i.f.[a] 2 years No findings [43]
Dialkyldimethylammoniumchloride $\left[\begin{smallmatrix} R & CH_3 \\ & N \\ R & CH_3 \end{smallmatrix}\right]^{(+)}$ $Cl^{(-)}$	5.000 [44] Rats 1.000 [44] tolerated by rabbits	Guinea pigs 1.000 mg/kg/day i.f.[a] 12 days No findings [45]

$R = C_{16-18}$　　　　　[a] In the food

in practice, however, a detoxification takes place as a result of the formation of neutral salts [82].

Even this brief statement of facts makes it clear that the surfactants used in detergents and cleaning agents today no longer represent a real problem in natural waters. Huber [90] summarises the observations made in the Federal Republic of Germany by stating that no deaths of fish are so far known to have occurred, if the products were properly used. In fact, because of the extensive biological treatment of the effluents, acute damage need not be feared.

Mammals

In addition to the establishment of the biological and ecotoxicological data, a knowledge of those properties of the surfactants which are relevant in human toxicology is an essential task. For this reason, an investigation of the effect on mammals is of particular importance for an estimate of the effect on man. Berth, Fischer and Gloxhuber [92] have given a review of the values for the acute and chronic toxicity of important surfactants (Table 11).

The authors state that the anionic surfactants, cationic surfactants and nonionic surfactants in common use today are acceptable from the point of view of human toxicology.

In chronic feed trials with linear alkylbenzenesulfonate, no indications of a carcinogenic action were found [93].

References

1. Kling, W.: Textilindustrie 69, 87 (1967)
2. Osteroth, D.: Dragoco-Rep. 26, 64 u. 78 (1979)
3. Bertsch, H.: Tenside 5, 185 (1968)
4. Großmann, H. in: Falbe, J., Hasserodt, U.: Katalysatoren, Tenside und Mineralöladditive, Georg Thieme-Verlag, Stuttgart, 1978, p. 123 ff.
5. Hintermaier, A.: Fette, Seifen, Anstrichmittel 59, 976 (1957)
6. FR. Pat. 766 903 (1934); Brit. Pat. 416 379 (1933); US. Pat. 2 220 099 (1934), I. G. Farbenind. (Günther, F., et al.)
7. US. Pat. 2 134 711 (1938), Nat. Aniline and Chem. Co. (Flett, L.H.)
8. Chem. Engng. News 22.05.1961, p. 56
9. Ströbele, R.: Chem.-Ing.-Techn. 36, 858 (1964)
10. Baumann, P.: IV. Int. Kongr. grenzflächen-akt. Stoffe, Brüssel 1964, Gordon and Breach, London 1968, Vol. 1, p. 1
11. Lewis, A.H.: US. Pat. 2 631 980 (1949), California-Res., Chem. Abstr. 47, 5706 a (1953)
12. Wulf, H.D., Böhm-Gössl, T.H., Rohrschneider, L.: Fette, Seifen, Anstrichmittel 69, 32 (1967)
13. Hirsch, E.: Vom Wasser 30, 249 (1964)
14. Bundesgesetzblatt, part 1, Nr. 72, 1653 (1961), Nr. 49, 698 (1962)
15. Kuhnen, L.: Chemie-Anlagen + Verfahren (CAV) 1978 Dec. S. 29
16. US-Pat 2 046 090 (1933) (Reed, C.F., Horn, C.L.)
17. Orthner, L.: Angew. Chem. 62, 302 (1950); Rösinger, S.: Chem.-Ing.-Techn. 42, 1236 (1970)
18. Püschel, F.: Tenside 4, 287 (1967); Kaiser, S., Püschel, F.: Chem. Ber. 97, 2926 (1964); Baumann, H., Stein, W., Voss, M.: Fette, Seifen, Anstrichmittel 72, 247 (1970)
19. US-Pat 3 676 523 (1972), Shell
20. Domagk, G.: Dtsch. Med. Wochenschr. 61, 829 (1935)
21. Rittmeister, W.: Melliand Textilber. 48, 1224 (1967)
22. Roelen, O.: Angew. Chem. 60, 62 (1948); DBP 931 405 (1939) Chem. Verwertungsges. Oberhausen m.b.H. (Landgraf, A., Roelen, O.)

23. Ziegler, K. et al.: Angew. Chem. *67*, 424, 425 (1955), DAS 1 014 088 (1954) (Ziegler, K.); Matson, T.P.: Soap and Chem. Spec. *39*, (11), 52–54, 91, 95–100 (1963)
24. Marti, B.: Seifen. Öle, Fette, Wachse *93*, 251 (1967)
25. Rudling, L., Solyom, P.: Water Res. *8*, 115 (1974); Fischer, W.K.: Tenside Detergents *8*, 182 (1971)
26. Schönfeld, N.: Surface Active Ethylen Oxid Adducts, Pergamon Press, Braunschweig, 1969, p. 69 ff.; Schick, M.: Nionic Surfactants, Marcel Dekker, New York, 1967, p. 149 ff.
27. Großmann, H.: Fette, Seifen, Anstrichmittel *74*, 58 (1972)
28. Malkemus, J.D.: J. Amer. Oil Chemists' Soc. *33*, 571 (1956)
29. US-Pat. 2 089 212 (1936) (Kritschewski, W.)
30. Smolka, J.R. in: Schick, M.: Nonionic Surfactants Marcel Dekker, New York, 1967, p. 300 ff., US-Pat. 2 674 619 (1953) Wyandotte Chem. Corp. (Lundstedt, L.G.)
31. Schmitz, A.: Fette, Seifen, Anstrichmittel *55*, 10 (1953); Schmitz, A.: III. Int. Kongr. grenzfl. akt. Stoffe, Köln 1960, Universitätsdruckerei Mainz 1963, B IV p. 264; Cramer, G.: Fette, Seifen, Anstrichmittel *60*, 35 (1958); Schöne, M.: Anfänge der Firma Henkel in Aachen und Düsseldorf, Werksarchiv Henkel GmbH, H. 5/6, 1973, Düsseldorf; Koch, P.A.: Ges. Textilind. *60*, 797 (1958); Weber, R.: Seifen, Öle, Fette, Wachse *95*, 885 (1969); Sinner, H.: Waschen mit Haushaltswaschmaschinen, Haus + Heim-Verlag, Hamburg, 1960
32. FRG, Stat. Bundesamt Wiesbaden: Industr. Produktion, Reihe 3 (1974), Verlag W. Kohlhammer, München; Seifen, Öle, Fette, Wachse *106*, 129 (1980); J. Amer. Oil Chemists' Soc. *57*, 282 A (1980); Krings, P., Harder, H., Weber, P.: Waschmittelchemie, Hüthig Verlag, Heidelberg, 1976, p. 9; Krings, P.: Fette, Seifen, Anstrichmittel *76*, (1974) 116; DBP 1056 316 (1966) Procter u. Gamble; DAS 1080 250 (1960) Procter u. Gamble; Kling, W.: Münchener Beitr. Abwasser-, Fischerei- u. Flußbiologie *12*, 38 (1965); Milster, H.: Seifen, Öle, Fette, Wachse *94*, 591 (1968); Berth, P., Jakobi, G., Schmadel, E.: Chemiker Ztg. *95*, 548 (1971); Kling, W.: Melliand Textilber. *46*, 957 (1965); Werdelmann, B.: Seifen, Öle, Fette, Wachse *83*, 123 (1957); Gilbert, A.H.: Detergent Age *4*, (7) (1967) 30; Murray, L.T.: J. Amer. Oil Chemists' Soc. *45*, 493 (1968); Werdelmann, B.: Soap, Cosmetics, Chem. Spezialit. *50*, (3) (1974) 36; Bloching, H.: Waschmittelchemie, Hüthig-Verlag, Heidelberg, 1976, p. 137; DRP 283 923 (1915) (O. Röhm); Hoogerheide, J.C.: Fette, Seifen, Anstrichmittel *70*, 743 (1968); Kretschmann, J.: Hauswirtschaft u. Wiss. *15*, (1968) 201; Kretschmann, J.: Tenside Detergents *13*, 5 (1975)
33. Barth, H., et al.: Fette, Seifen, Anstrichmittel *68*, 48 (1966)
34. Jakobi, G., Schwuger, M.J.: Chemiker Ztg. *99*, 182 (1975); Henning, K., Merkenich, K. in: Falbe, J., Hasserodt, U.: Katalysatoren, Tenside, Mineralöladditive, Georg Thieme-Verlag, Stuttgart 1978, p. 152
35. Berth, P., Jakobi, G., Schmadel, E., Schwuger, M.J., Krauch, C.M.: Angew. Chem. *87*, 115 (1975)
36. Heins, A.: Seifen, Öle, Fette, Wachse *102*, 576 (1976); Henning, K.: Tenside Detergents *13*, 208 (1976); Schwuger, M., Smolka, H.G., Kurzendörfer, C.P.: ibid. *13*, 305 (1976)
37. Leschber, R., Au, I.: Tenside Detergents *16*, 212 (1979); Seifen, Öle, Fette, Wachse *105*, 337 (1979), *106*, 126 (1980)
38. Großmann, H.: Reiniger u. Wäscher *30* (76), 9, 25 (1977); Fischer, K.: Melliand Textilber. *59*, 487, 582, 659 (1978)
39. Reng, A., Faber, R., Quack, J. in: Stache, H.: Tensidtaschenbuch, Carl-Hanser, München 1979, p. 153
40. Griffin, W.C.: J. Soc. Cosmet. Chem. *1*, 311 (1949)
41. Ludwig, K.G., Gackenheimer, W.C.: Fette, Seifen, Anstrichmittel *70*, 567 (1968); Diannisi, G.F.: Tenside *2*, 40 (1965)
42. Schultze, G.R.: III. Int. Kongr. grenzfl.-akt. Stoffe, Köln, 1960, Universitätsdruckerei Mainz 1963, Vol. 4, p. 617
43. Schuller, H.: Tenside *2*, 83 (1964); Heusch, R.: Tenside Detergents *12*, 3 (1975)
44. Taube, P.: V. Int. Kongr. grenzfl.-akt. Stoffe, Barcelona, 1968, Vol. 2, p. 387; Dambacher, F.: Tenside *5*, 24 (1968)
45. McKee, R.H.: Ind. Engng. Chem. *38*, 382 (1946)
46. US-Pat. 2 768 894, GAF
47. Fiedler, H.P. in: Falbe, J., Hasserodt, U.: Katalysatoren, Tenside u. Mineralöladditive, Georg-Thieme, Stuttgart 1978, p. 178
48. Heusch, R., Niessen, H. ibid., p. 182
49. Verband Chem. Ind., Frankfurt: Physikalische u. chem. Reinigungsverfahren (1975)

50. Malz, F.: III. Internat. Kongr. grenzflächenaktive Stoffe 12.–17. Sept. 1960, Köln-Mainz, Vol. 3
51. FRG, Bundesgesetzblatt, part I, Nr. 128 (1976), p 3017–3032
52. Krüger, R.: Erdöl und Kohle. Erdgas. Petrochemie. *16*, 379 (1963)
53. Dinkloh, L., Au, I.: Tenside-Detergents *17*, 236 (1980)
54. FRG, Bundesgesetzblatt, part I, Nr. 72 (1961), p. 1653
55. Ibid. Nr. 49 (1962), p. 698
56. Pollution by Detergents, OECD, Paris, 1971
57. Amtsblatt Europ. Gemeinschaften, 17.12.1973, Nr. L 347/51–52
58. Ibid. Nr. L 347/53–63
59. FRG, Bundesgesetzblatt, part I, Nr. 100 (1975), p. 2255
60. Proposed Method for the Determination of the Biodegradability of Surfactants Used in Synthetic Detergents, OECD, Paris, 1976
61. FRG, Bundesgesetzblatt, part I, Nr. 9 (1977), p. 244
62. Ibid. Nr. 128 (1976), p. 3017
63. Wickbold, R.: Analytik der Tenside, Chemische Werke Hüls AG, Marl, 1976
64. Kunkel, E.: Tenside-Detergents *17*, 247 (1980)
65. Wickbold, R.: ibid. *8*, 140 (1971)
66. Deutsche Einheitsverfahren zur Wasser-, Abwasser- und Schlammanalyse. Verlag Chemie, Weinheim, DIN 38 409, part 23
67. Kunkel, E.: Disulfinblaumethode, hüls-Monografie: Die Analytik der Tenside, S. 102, 1976
68. Michelsen, E.R.: Tenside-Detergents, *15*, 169 (1978)
69. Waters, J., Kupfer, W.: Analyt. Chim. Acta *85*, 241 (1976)
70. Bock, K.J., Schöberl, P. in: Chwala, A., Anger, V.: Handb. Textilhilfsmittel, Weinheim, 1977, p. 1045
71. Bock, K.J., Schöberl, P. in: Falbe, J., Hassenrodt, U.: Katalysatoren, Tenside and Mineralöladditive, Stuttgart, 1978, p. 213
72. Schöberl, P., Bock, K.J.: Tenside-Detergents *17*, 262 (1980)
73. Jendreyko, H., Bock, K.J.: Gas- und Wasserfach *103*, 615 (1962)
74. Krüger, R.: Fette, Seifen, Anstrichmittel *66*, 217 (1964)
75. Swisher, R.D.: Surfactant Biodegradation, Surfactant Sci. Ser., Vol. 3, Marcel Dekker, New York 1970
76. Steber, J.: Tenside-Detergents *16*, 5 (1979)
77. Heinz, H.J., Fischer, W.K.: Fette, Seifen, Anstrichmittel *66*, 270 (1962)
78. Husmann, W., Malz, F., Jendreyko, H.: Forschungsber. Nordrhein-Westf., Nr. 1153. Westdeutscher Verlag, Köln, 1963
79. Spohn, H.: Tenside-Detergents *1*, 18 (1964); Spohn, H., Fischer, W.K.: ibid. *1*, 87 (1964); *4*, 241 (1967)
80. Bock, K.J., Wickbold, R.: Vom Wasser *33*, 243 (1966)
81. May, A., Neufahrt, A.: Tenside-Detergents *13*, 65 (1976)
82. Huber, L.: Münchener Beitr. Abwasser-, Fischerei- u. Flußbiologie *31*, 203 (1979)
83. Fischer, W.K.: Tenside-Detergents *17*, 250 (1980)
84. Fischer, W.K., Winkler, K.: Vom Wasser *47*, 81 (1976)
85. Waters, J.: ibid. *47*, 132 (1976)
86. Hirsch, E.: Fette, Seifen, Anstrichmittel *65*, 814 (1963); Vom Wasser *30*, 249 (1963); Divo, C.A.: 12th World Congr. Internat. Soc. Fat Res., Milano, Sept. 1974
87. Bock, K.J.: Arch. Fischereiwiss. *17*, 68 (1966)
88. Niemitz, W., Pestlin, W.: Städtehygiene *13*, 231 (1962); Herbert, D.W.M., Elkins, G.H.J., Mann, E.T., Hemens, J.: Water and Waste Treatment J. *6*, 394 (1957)
89. Schöberl, P., Kunkel, E.: Tenside-Detergents *14*, 293 (1977)
90. Huber, L.: ibid. *17*, 267 (1980)
91. Alabaster, J.S.: J. Am. Oil Chemists Soc. *55*, 181 (1978)
92. Berth, P., Fischer, W.K., Gloxhuber, Chr.: Tenside-Detergents *9*, 260 (1972)
93. Bornmann, G., Loeser, A.: Z. Lebensmittel-Untersuch. und -Forsch. *118*, 51 (1962)

Subject Index

References of Part A are marked A, References of Part B are marked B.

Example: PCB   A: 164; B: 89

The Handbook of Environmental Chemistry

Editor: O. Hutzinger

This handbook is the first advanced level compendium of environmental chemistry to appear to date. It covers the chemistry and physical behavior of compounds in the environment. Under the editorship of Prof. O. Hutzinger, director of the Laboratory of Environmental and Toxicological Chemistry at the University of Amsterdam, 37 international specialists have contributed to the first three volumes.
For a rapid publication of the material each volume is divided into two parts. Each volume contains a subject index.

The Handbook of Environmental Chemistry is a critical and complete outline of our present knowledge in this field and will prove invaluable to environmental scientists, biologists, chemists (biochemists, agricultural and analytical chemists), medical scientists, occupational and environmental hygienists, research geologists, and meteorologists, and industry and administrative bodies.

Springer-Verlag
Berlin
Heidelberg
NewYork

Volume 1 (in 2 parts)
Part A

The Natural Environment and the Biogeochemical Cycles

With contributions by numerous experts
1980. 54 figures. XV, 258 pages
ISBN 3-540-09688-4

Contents:
The Atmosphere. – The Hydrosphere. – Chemical Oceanography. – Chemical Aspects of Soil. – The Oxygen Cycle. – The Sulfur Cycle. – The Phosphorus Cycle. – Metal Cycles and Biological Methylation. – Natural Organohalogen Compounds. – Subject Index.

Volume 2 (in 2 parts)
Part A

Reactions and Processes

With contributions by numerous experts
1980. 66 figures, 27 tables. XVIII, 307 pages
ISBN 3-540-09689-2

Contents:
Transport and Transformation of Chemicals: A Perspective. – Transport Processes in Air. – Solubility, Partition Coefficients, Volatility, and Evaporation Rates. – Adsorption Processes in Soil. – Sedimentation Processes in the Sea. – Chemical and Photo Oxidation. – Atmospheric Photochemistry. – Photochemistry at Surfaces and Interphases. – Microbial Metabolism. – Plant Uptake, Transport and Metabolism. – Metabolism and Distribution by Aquatic Animals. – Laboratory Microecosystems. – Reaction Types in the Environment. – Subject Index.

Volume 3 (in 2 parts)
Part A

Anthropogenic Compounds

With contributions by numerous experts
1980. 61 figures, 73 tables. XIII, 274 pages
ISBN 3-540-09690-6

Contents:
Mercury. – Cadmium. – Polycyclic Aromatic and Heteroaromatic Hydrocarbons. – Fluorocarbons. – Chlorinated Paraffins. – Chloroaromatic Compounds Containing Oxygen. – Organic Dyes and Pigments. – Inorganic Pigments. – Radioactive Substances. – Subject Index.